BUILDING FUTURES

BUILDING FUTURES

Technology, Ecology, and Architectural Practice

WILEY

Richard Garber

Copyright © 2023 by John Wiley & Sons, Inc. All rights reserved.

Published by John Wiley & Sons, Inc., Hoboken, New Jersey.
Published simultaneously in Canada.

No part of this publication may be reproduced, stored in a retrieval system, or transmitted in any form or by any means, electronic, mechanical, photocopying, recording, scanning, or otherwise, except as permitted under Section 107 or 108 of the 1976 United States Copyright Act, without either the prior written permission of the Publisher, or authorization through payment of the appropriate per-copy fee to the Copyright Clearance Center, Inc., 222 Rosewood Drive, Danvers, MA 01923, (978) 750-8400, fax (978) 750-4470, or on the web at www.copyright.com. Requests to the Publisher for permission should be addressed to the Permissions Department, John Wiley & Sons, Inc., 111 River Street, Hoboken, NJ 07030, (201) 748-6011, fax (201) 748-6008, or online at http://www.wiley.com/go/permission.

Trademarks: Wiley and the Wiley logo are trademarks or registered trademarks of John Wiley & Sons, Inc. and/or its affiliates in the United States and other countries and may not be used without written permission. All other trademarks are the property of their respective owners. John Wiley & Sons, Inc. is not associated with any product or vendor mentioned in this book.

Limit of Liability/Disclaimer of Warranty: While the publisher and author have used their best efforts in preparing this book, they make no representations or warranties with respect to the accuracy or completeness of the contents of this book and specifically disclaim any implied warranties of merchantability or fitness for a particular purpose. No warranty may be created or extended by sales representatives or written sales materials. The advice and strategies contained herein may not be suitable for your situation. You should consult with a professional where appropriate. Further, readers should be aware that websites listed in this work may have changed or disappeared between when this work was written and when it is read. Neither the publisher nor authors shall be liable for any loss of profit or any other commercial damages, including but not limited to special, incidental, consequential, or other damages.

For general information on our other products and services or for technical support, please contact our Customer Care Department within the United States at (800) 762-2974, outside the United States at (317) 572-3993 or fax (317) 572-4002.

Wiley also publishes its books in a variety of electronic formats. Some content that appears in print may not be available in electronic formats. For more information about Wiley products, visit our web site at www.wiley.com.

Library of Congress Cataloging-in-Publication Data is applied for

Paperback ISBN: 9781119829218

Cover Image: Courtesy of GRO Architects, PLLC
Cover Design: Wiley

CONTENTS

xi   **FOREWORD**

2   **INTRODUCTION: BUILDING FUTURES**

5   **PART 1**

6   **CHAPTER 1**   On Technology I

24   **CHAPTER 2**   Morphosis's Immaterial Moments in the Making of *Things*

46   **CHAPTER 3**   On Technology II

58   **CHAPTER 4**   UNStudio's Future Lifecycles

81   **PART 2**

82   **CHAPTER 5**   On Ecology I

94   **CHAPTER 6**   Zaha Hadid's Circular Economy

118   **CHAPTER 7**   On Ecology II

132   **CHAPTER 8**   Winka Dubbeldam's Synthetic Natures

153   **PART 3**

154   **CHAPTER 9**   On Construction I

166   **CHAPTER 10**   Componibile, by Remote Control, GRO Architects

178   **CHAPTER 11**   On Construction II

198   **CHAPTER 12**   On Practice

206   **CHAPTER 13**   Assembly OSM's Modular Platforms and Digital Twins

224   **CHAPTER 14**   A Practical Synopsis

235   Index

For Reed, Daryn, and other inheritors of our world.

FOREWORD
ROBERT STUART-SMITH

Architecture is a material response to a diverse array of social, economic, environmental, geopolitical, and discursive conditions. Its means, however, is inherently tied to technological progress arising from both within and outside of the profession. Most notably today, the profession's highly collaborative activities are framed significantly by software that facilitate the conception, communication, and delivery of architectural works. Yet, following creative explorations into digital design in the 1990s and the global adoption of building information modeling (BIM) in practices during the last two decades, the relation of architecture to software is, for many of us, simply a moment in history. Despite its ever-present role in practice, it is seldom a topic of discussion. In particular, BIM, is commonly associated with software programs developed by a handful of companies that has extended 2D and 3D computer-aided design (CAD) into information-rich 3D models. The profession's approach to BIM, however, does not need to remain in this mindset that is focused solely on the partial automation of construction documentation, as there is much more at stake.

The National Institute for Building Standards defines BIM as "a digital representation of physical and functional characteristics of a facility"[1] that "serves as a shared knowledge resource for information about a facility forming a reliable basis for decisions during its lifecycle from inception onwards."[2] This description is quite similar to that of the *digital twin*, a term coined by Michael Grieves in 2002 but first practiced by NASA during the 1960s. In a 2010 draft report, NASA describes the digital twin as "an integrated multi-physics, multi-scale, probabilistic simulation of a vehicle or system that uses the best available physical models, sensor updates, fleet history, etc., to mirror the life of its flying twin."[3] For NASA, digital twins allow realistic design and problem-solving activities to take place rapidly, with low risk of failure compared to more historically abstract approaches.

This book argues that a forward-looking approach to architectural workflows can also shift BIM toward greater degrees of *digital twin* thinking, and critically and creatively investigates where this could lead.

Today's digital twins can embed and associate data on 3D objects, enabling architects to correlate a vast network of products, manufacturing processes, and material specifications. In other industries, digital twins are already used to assess the workings of several interrelated mechanical parts, or to build entire simulated environments to model scenarios for technologies that must be future-proof (e.g. cell phone reception relative to the placement of radio transmitter towers). In *Building Futures*, Richard suggests we can go much further, drawing our attention to the *Anthropocene* through the lens of Timothy Morton's concept of *hyperobjects*.[4] The world's human-generated climate crisis is presented as one such hyperobject that exceeds both our lifespan and full comprehension, suggesting there must be greater architectural accountability and agency to address it.

The book prompts us to consider simulating events where architecture and architects could mitigate, redirect, or develop contingencies in relation to the environment, flows of material and capital, and other "things" that operate from the immediate, through to almost the geological timescales Manuel Delanda portrays in *A Thousand Years of Nonlinear History*[5] where time is described as a history of flows of geology, disease, and economy, amongst others. This is extended into the broader perspective of *posthumanism*, which Cary Wolfe describes as a "historical moment in which the decentering of the human by its imbrication in technical, medical, informatic, and economic networks is increasingly impossible to ignore."[6] Posthuman approaches to design are advocated for, that recognize there are other species and things that need to be designed for and with, challenging the fact that most architecture is anthropocentric. The book suggests that digital twin simulation and modeling could reframe ideas and the performances of buildings and their respective parts within broader ecological timescales, and calls for more holistic thinking around building design. But is *holistic* the appropriate term? Perhaps not.

The *speculative realist* philosopher, Levi Bryant, advocates for a "flat ontology"[7] (flattened hierarchy), where the human is considered an object, the same as any other object or thing, able to influence and be influenced by any other thing. In Bryant's flat ontology there is no world, container, or meta object that represents completeness or unity. Such a perspective aligns with the book's position, challenging classical notions of architecture's part-to-whole relationships, distancing itself from ideas of composition toward those of assemblage, and questioning whether a building is actually ever a whole or whether its parts are related to, or contribute to the existence of other parts within a flat ontology.

The call to rethink architecture's relation to other objects extends beyond those of object-orientated ontology, championed by Bryant, as well as Graham Harman and Ian Bogost, to the altogether different matter of object-orientated programming.[8] Richard calls into question how architects engage with software, and whether we are constrained by the commercially supported tools we use. Through a series of case studies, the book illustrates the degree to which software is customized by designers from the crafting of bespoke BIM and other workflows by those without programming experience, to scripting customization within 3D modeling software, and beyond, into more serious in-house software development. While software does indeed confine and limit capabilities, the case studies demonstrate that several offices' custom workflows for individual projects have enabled previously infeasible design options to be economically and practically viable within workable design and delivery approaches.

Programming is one of many topics that are used to reframe the scope and agency of the architect, alongside architects' almost direct engagement in digital design-to-production activities such as CNC manufacturing. Increases in productivity and optimizations in project team size and structure resulting from BIM workflows are also touched on. The historically conflicting ideologies of architects, developers, and construction project managers are questioned, suggesting that their disparate interests can align, or at least be resolved as additional criteria within architectural outcomes. If one views architecture as polyvalent and incorporating diverse performative, aesthetic, and constructive criteria, among others, then why not? While BIM is considered the purview of those involved in construction documentation, the book argues that such methods can be extended into feasibility and schematic work, also engaging with real estate development, to further align the interests of diverse parties involved in project initiation and realization. This positive outlook on the increasingly specialized nature of the construction industry is perhaps most notable in discussions on modular construction that is expanding the scope of off-site prefabricated building solutions.

While others see such turnkey design-build companies as a threat to architecture, such as the late Katerra that attempted to fuse architecture and construction within one enterprise, the book takes a different view. Rather than casting the rise of modular construction as a diktat thrust upon architects to build tightly to a predetermined set of specifications, dialogue is demonstrated where architects engage more directly with fabrications during design than has been historically possible, providing more scope for design customization to be jointly developed by the fabricator and designer in partnership, potentially resulting in reduced cost and risk.

Case study projects from several practices are also discussed, where architects have expanded design agency in diverse ways. Given that these developments have been decades in the making, why write this book now? Relative to present and near-future developments, the timeliness of this publication could not be more impactful. First, there is environmental and socioeconomic urgency. Buildings have a substantial environmental footprint, accounting for 39% of energy, 40% of CO_2 emissions, and 40% of raw materials used each year.[9] There is a great need to develop design solutions that reduce this impact and slow climate change. Socioeconomically, many developed and developing economies cannot keep up with demand to adequately house their current populations, not to mention the 9 billion people expected to be alive by 2045 or the estimated 6 billion additional people who will be living in urban populations by 2050.[10] To meet these projections, the building sector needs to become far more productive. Although construction is currently the least productive manufacturing sector, there is hope. A UK government report determined that construction productivity could improve by 60 percent simply by undertaking at least 70 percent of activities off-site prior to construction, where factory-like conditions provide greater degrees of safety and control, and are not disrupted by weather or traffic.[11] Such productivity improvements have the potential to also generate reductions in the cost and time of building, supporting a more affordable architecture.

Government regulation and industry are embracing digital twin technologies at an unprecedented level while emerging technologies are extending the sphere of influence a digital twin can have. Since 2008, there are more things communicating with other things on Earth than people communicating with people. Often referred to as the Internet of Things (IOT), the number of devices or objects connected to the internet that can communicate with people and other objects is estimated to reach 75 billion by 2025.[12] Although typically associated with "smart" appliances such as TVs and fridges that have a computer and internet connection, IOT technologies have given rise to numerous distributed sensor and data-capture devices that track building construction progress or monitor building operations, such as the functioning of MEP systems or structural failure in bridges. Beyond inert IOT objects, there are now commercial companies offering a range of mobile robot systems that can undertake construction site survey mapping,[13] simple building tasks,[14] or be deployed for infrastructural repair tasks.[15] While companies offering these services currently operate in silos, LivingPlanIT is developing an urban operating system (urban OS) that promises to be the glue between everything in the built environment – the ultimate digital twin.[16]

These developments are taking place at the commencement of the Fourth Industrial Revolution (Industry 4.0), which extends Industry 3.0 information and technology systems into autonomous manufacturing and cyber-physical systems. Industry 4.0 promises to enable greater levels of automation and user customization in the physical world through digital technologies. Klaus Schwab, economist and founder of the World Economic Forum, describes Industry 4.0 as having four main physical outcomes in the short term: autonomous vehicles, 3D printing, advanced robotics, and new materials.[17] All of these developments are already influencing building design and construction activities, yet their impact is likely to increase. As they do, BIM digital twin design workflows will be able to more fundamentally connect to these transformations to material and product supply chains, manufacturing, and the delivery of buildings with greater awareness of product life cycles, environmental impact, and the financial and construction risk implications of design decisions. Such digital interconnectivity can support supply chains operating both more globally, and more locally. Industry 4.0 technologies might also enable economical means of distributed manufacturing that could support more locally sourced materials or manufacturing methods.

In the University of Pennsylvania and University College London's Autonomous Manufacturing Lab (AML) we have been developing a multi-robot autonomous manufacturing software framework to support a distributed, adaptive approach to off-site building prefabrication and on-site construction data collection. By connecting heterogeneous teams of robots (mobile ground robots, aerial robots, industrial robot arms, track and gantry systems) to a digital twin model, the manufacture and assembly of a building or building part is able to be decomposed into several discrete operations that can be autonomously selected and executed by robots who work collectively. While this research is relatively nascent in vision and capability, it is being developed in collaboration with industry and the UK government, with significant input from construction companies, engineering firms, and community groups.[18] From this work, it is clear that there is an industry need, and significant gains to be made in human safety, productivity, flexibility, and improved precision, quality assurance, predictability, and risk mitigation. This research offers a glimpse into how valuable digital twin models will become during production, yet they will also have an increasingly significant impact on design.

From a design perspective, Industry 4.0 will facilitate a shift from the dominance of Fordist mass production to a larger amount of post-Fordist mass-customized designs. Fordist production gave rise to modular prefabricated building components that architecturally allowed buildings to be expressed as an assemblage of identical large-scale parts. Although these techniques remain of practical importance, post-Fordist capabilities create space for industrially scalable bespoke production, that lends itself to more variation in the geometric definition of a building's respective parts, their collective expression, and greater design variation between buildings. Designs can become more unique, site-specific, and user-customized, tailored to specific programmatic or climatic conditions. BIM will be the space in which these possibilities are explored, tested, and realized.

Beyond any one specific design capability, Industry 4.0 will provide the architect with increased design flexibility. Some Industry 4.0 technologies, such as additive manufacturing, are relatively design-agnostic, with costs solely related to material volume and build time. Architects can therefore gain aesthetic freedoms providing they engage with the environmental and economic consequences of their aesthetic decisions. It is hoped this will prompt architects to rethink a vast array of relationships and activities around design conception, fabrication, transportation, assembly, building occupation and use, disassembly, reuse, and recycling, among others.

Industry 4.0's reliance on cyber-physical systems also gives rise to a convergence between software and hardware that questions the very nature and modus operandi of architecture itself. Architectural design intent can now be embodied within responsive, adaptive systems, some unseen (such as AI-driven MEP systems) and some aesthetic, user-customizable, or kinetic. Every aspect of a building has the potential to be an IOT part, including a building's own parts. Due to this, an architectural design brief might no longer be so anthropocentric. For instance, the retailer Ocado operates warehouses whose primary occupants are robots. In the journal *Science Robotics*, together with Vijay Pawar and Peter Scully, I argued that architecture itself might soon be thought of as a robotic and autonomous system, with the built environment or buildings themselves comprising an ecology of robot systems.[19]

Governmental, social, and corporate interest toward some form of "metaverse" is also increasing, which, together with web 3.0, blockchain, and cryptocurrency infrastructure, will enable unprecedented connectivity between physical and virtual information, activities, and networks. A BIM digital twin in this context will profoundly impact the way we conceptualize, collaborate, and engage with stakeholders and realize projects.

Dialogue might involve direct user customization or be indirect through feedback from sensory systems. In both cases, a BIM digital twin will operate center stage.

We are at the beginning of a major shift – not only in the means we undertake building design and construction but also in the way we can perceive and initiate architectural agency. It is an exciting time to be in practice, as new connections and agencies can be established to support a more inclusive, ambitious, and impactful architecture. At the same time, a more technological architecture could in some sense become more down-to-earth, extending dialogue to greater degrees with end users, and engaging more directly with many of the issues raised in this book. This book does not champion the status quo of today's BIM approaches, but recasts BIM as a platform in which to further extend our critical and creative thinking – to ask, where do we go from here? The plural nature of the title *Building Futures* implies that there is more than one path forward. In the chapters that follow, possible building futures are explored within speculations on technology, ecology, construction, and practice that are also supported by chapters devoted to a series of high-interest case study projects. These elevate BIM to something beyond its colloquial trivial meaning, pointing toward ideas, concerns, and opportunities that should be at the forefront of every architect, academic, researcher and historian's mind. Curiously, Richard's first chapter commences with a quote from *Tenet* – perhaps one of the most mind-blowing science-fiction films of the decade. As does *Tenet*, the book stitches together events from the recent past in order to influence our future trajectory. *Building Futures* thus really speaks to the one moment that can always change what lies ahead – the here and now.[20]

NOTES

1 National Institute of Building Sciences, *National BIM Guide for Owners*, (2017), p. 12, https://www.nibs.org/reports/national-bim-guide-owners.
2 Ibid, p. 30.
3 E. H. Glaessgen and D.S. Stargel, "The Digital Twin Paradigm for Future NASA and U.S. Air Force Vehicles," 53rd *Structures, Structural Dynamics, and Materials Conference: Special Session on the Digital Twin* (American Institute of Aeronautics and Astronautics, 2012), p. 7.
4 Tatiana Morton, "Hyperobjects: Philosophy and Ecology after the End of the World." *International Journal of Environmental Studies*, 76(3): 523–524, doi: 10.1080/00207233.2018.1536436
5 Manuel de Landa, *A Thousand Years of Nonlinear History* (New York: Zone Books, 1997).
6 Cary Wolfe, *What Is Posthumanism?* (Minneapolis: Univ. of Minnesota Press, 2009). xv.

7 Levi Bryant, *The Democracy of Objects* (Open Humanities Press, 2011), p. 73.
8 Graham Harman, *Object-Oriented Ontology: A New Theory of Everything* (Penguin Books Limited, 2018).
9 David Bergman. *Sustainable Design : A Critical Guide*, 1st ed. (New York: Princeton Architectural Press, 2012). pages 15, 24–25, 62.
10 United Nations Department of Economic and Social Affairs, World Urbanization Prospects: The 2014 Revision, New York, 2015. See https://population.un.org/wup/publications/files/wup2014-report.pdf
11 Science and Technology Select Committee, *Off-Site Manufacture for Construction: Building for Change* (London: House of Lords, 2018) https://publications.parliament.uk/pa/ld201719/ldselect/ldsctech/169/169.pdf.
12 "Statista Research Department, *Number of IoT devices 2015–2025*," accessed January 7, 2022, https://www.statista.com/statistics/471264/iot-number-of-connected-devices-worldwide/
13 "Trimble, Spot® — Trimble's Robotic Autonomous Scanning Solution," updated 2022, https://fieldtech.trimble.com/en/product/spot
14 "Dusty Robotics, *Construction Robots: BIM-Driven Layout*," updated 2023, https://www.dustyrobotics.com/; Hilti USA, "Hilti Unveils BIM-Enabled Construction Jobsite Robot: Construction Automation Solution to Help Contractors Tackle Productivity, Safety, and Labor Shortage Challenges on Jobsites," updated October 28, 2020, https://www.hilti.com/content/hilti/W1/US/en/company/news/press-releases/2020-construction-jobsite-robot.html; nLink AS. *"nLink Robotics,"* updated 2021, https://www.nlinkrobotics.com/about-nlink-robotic-company.
15 Sophie Burkholder, "Gecko Robotics Closed on a $73M Series C in Pittsburgh's Biggest Deal of 2022," March 3, 2022, https://technical.ly/startups/gecko-robotics-series-c/; Eugene Demaitre, "Infrastructure Robots Rise to Meet America's Urgent Needs," *Robotics Business Review*, July 4, 2018, https://www.roboticsbusinessreview com/construction/infrastructure-robots-meet-urgent-needs/
16 Living-PlanIT AG, "Living-PlanIT AG," updated 2020, http://www.living-planit.com/
17 Klaus Schwab, *The Fourth Industrial Revolution* (London, England: Portfolio Penguin, 2017).
18 EPSRC, "EPSRC GoW: Applied Off-site and On-site Collective Multi-Robot Autonomous Building Manufacturing," retrieved

November 13, 2022, https://gow.epsrc.ukri.org/NGBOViewGrant.aspx?GrantRef=EP/S031464/1

19 Pawar VM, Stuart-Smith R, Scully P. Toward autonomous architecture: The convergence of digital design, robotics, and the built environment. Sci Robot. 2017 Apr 26;2(5):eaan3686. doi: 10.1126/scirobotics.aan3686. PMID: 33157895.

20 C. Nolan, *Tenet* (Warner Bros. August 26, 2020).

INTRODUCTION: BUILDING FUTURES

Building information modeling (BIM) has seen widespread use and adoption in the architecture, engineering, and construction (AEC) industry in recent years, with general agreement in both the applicability and use of BIM systems to support aspects of design, construction, and post-occupancy facilities management. Most efforts to enhance both standardization and efficiency have been coordinated by groups such as the US National Institute of Building Sciences (NIBS) and American Society of Civil Engineers (ASCE), which offer publications and events as well as consultation and advocacy in support of these goals. A perusal of information available suggests a forward-looking, if not slightly dated, attitude toward building information modeling, with slogans such as "the future starts with civil engineers" or ". . . buildings of tomorrow" or "what the industry sees today and predicts tomorrow."[1] Given the goal of expanding both the utility and territory of information modeling practices here, it is useful to understand how these tools are regarded by the industry.

Quoting the US National Institute of Building Sciences, which has been cited in some of my earlier work on this subject, building information modeling refers to the "use of the concepts and practices of open and interoperable information exchanges, emerging technologies, new business structures, and influencing the reengineering of processes in ways that dramatically reduce multiple forms of waste in the building industry." This definition, which existed as early as 2008,[2] clearly suggests that building information modeling should be treated as more than a software or set of digital tools but as a series of protocols with goals of achieving a more sustainable basis for the work that architects and our allied collaborators perform in the service of Earth.

Waste in this sense has more generally been measured by efficiency, as in *waste of time* or *waste of money,* than it is with the necessary goal of reducing excess in the form of carbon emissions or construction refuse; and much of the work being focused on now with respect to standardization involves specific relationships between design, the (increasingly) virtual data generated to support the construction of a building and that construction (labor) itself.

These efforts are noble, and by virtue bring architects closer to both the building site and the very craft of building – places and actions from which we have been long since removed by both contract and practice. The rapid advancement of technology has not only created intense competition among contracting firms, but also the methodologies used in building construction – the *means and methods* we as architects have been taught to avoid. Such change is welcomed by many and is slowly being adopted as larger and more complex aspects of the industry, which include material procurement and supply chain positioning against a backdrop of global demand, are exposed. The intensity of these pressures, following a global pandemic and increasing awareness of – and involvement with – nature events precipitated by global effects of climate change, finally allow for a broadening of what precisely building information modeling is, and how we architects and designers access its potentials, given these advancements in both computation as well as an expansion in the scope of architectural process.

Building information modeling as the process of design, via engagement of new digital protocols, has the capacity to both raise specific issues through *design research* and solve specific problems through *applied design*. This idea of exposing issues through a process of research has been a core aspect of the work of the architect and is a practice we should seek to duly expand through the role of both *modeling* and *models*, in the design process, ensuring we are at once not relegated simply as problem solvers, but also imparted with the ability to understand and engage the broad and complex problems of building. An information model is a virtual database that can accept, process, and simulate multiple constraints and inputs, including creative and economic flows of capital, and is the result of a design process. It is the result of creative work and a physical output of design and is therefore representative not only of design intent but also our very agency as architects.

I would like to thank GRO Architects and in particular Mia D'Alessandro, Huajie Ma, Jake Hamilton, Zeina Husayni, and Yifan Shi in the preparation of images for this book.

NOTES

1 Procore Global Report "Construction: The Next Five Years," https://www.procore.com/
2 "National BIM Standard Released," *Buildings*, April 1, 2018, https://www.buildings.com/articles/34678/national-bim-standard-released, accessed May 27, 2022.

PART 1

1 | ON TECHNOLOGY I

The Protagonist, "He can communicate with the future?"

Priya, "We all do, don't we? Email, credit cards, texts. Anything that goes into the record speaks directly to the future. The question is, can the future speak back?"
— Tenet, 2020

INFORMATION, OBJECTS, AND THE EXPANDED FIELD OF ARCHITECTURAL DESIGN

On a recent, albeit prepandemic, trip to Beijing, a colleague there described an entertaining, if not concerning, story about the use of software in the service of architectural design. Building projects in China are increasingly required to be completed using building information modeling (BIM) software so that a virtual model can be tendered as part of a final drawing submission – a process already broadly adopted in Europe and the United States; however, many Chinese-based firms are still in the process of adopting these technologies and are simply completing building design with traditional two-dimensional CAD systems and providing a three-dimensional model as an afterthought. As such, the model is neither "live" nor "connected" to the drawing set – it is in effect a stagnant set of virtual data.

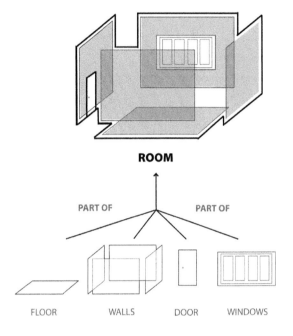

There has been a steady pressure to adopt BIM platform(s) rooted in a more *efficient* way of working. This has culminated with many architecture firms coming to the schools looking to hire recent graduates who simply "know BIM," as they either have a project whose deliverables require a model, or there is a general sense that other firms have invested in the technology so why shouldn't they. At some level, this is not all bad – many firms have created positions for BIM managers. However, there is a contradiction that we must confront: while most young architectural designers are versed in software, the digital tools utilized to support the architectural design process, they have less experience in building or, more specifically, in how such tools relate to construction. This is a contradiction that has serious implications in the creative use of these tools and how we conceptualize their use in establishing new workflows or processes in the service of design.

There seems to be disparate trends within the schema of modeling tools, which continue to become more sophisticated without measurable impact in the building industry, perhaps pointing to a lack of conceptualization of the tools themselves and their use in the service of architectural design. Utilization of these tools directly concerns our agency as architects and our relationship to building, yet also demonstrates an instrumental, if not consumerist, interest in the adoption of a new technology – BIM. Such a position privileges an efficiency of working methods as opposed to the engagement of information models as a new suite of modeling and simulation tools that support novel exploration and creative probing of design problems. Still, it is the "unmatched potential of technical drawing to refer to the material world that make it so extraordinarily effective in the representation of architecture.[1]"

New modeling operations have been continuously refined in both a technical and speculative way through more specific integrations with geometry and process. Equally important in this process has been the relationship between design and abstraction. Though abstraction plays a critical role in design, specific aspects of BIM have sought to reduce abstraction, which is sometimes incorrectly understood as "imprecision." A specific role of the digital twin concept is to minimize abstraction. To better understand this, abstraction as it relates to both preliminary design operations and construction phase services should be more precisely defined. Abstraction is not *vagueness*; in a design sense, it involves the isolation of certain design variables so they can be better understood relationally as they support the interconnectedness of architectural objects and allow the designer to better understand part-to-whole relationships. When understood this way, abstraction becomes an equally important tactic when making more downstream design decisions, where integrated project workflows begin to solely focus on efficiency.

Abstraction at the stage of building actualization can be a powerful operation when utilized within a virtual environment in understanding specific relationships between building components. Whether this is a relationship of a structural member to an architectural feature or a plumbing or gas riser to a wall cavity or mechanical shaft, abstraction in this sense involves isolation within a virtual construct to better understand a relationship between *things*. Such operations are increasingly occurring in virtual or augmented reality environments that allow teams to isolate discrete components to study how they are implemented during, and relate to other objects through, construction.

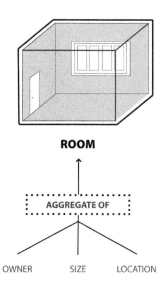

Figure 1.1 Meronyms and Holonyms, 2022: Part-to-whole project relationships as understood through modeling technologies allow better visualization and convey understandings of how things go together. These relationships increasingly anticipate use and post-occupancy scenarios following the construction phase of a building. Interestingly, in object-oriented computing, an architectural example – a room – is used in establishing aggregate taxonomies.[2] The term *meronymic* is used in establishing the relationship between a meronym (part) and holonym (whole). A window is a meronym of a room, which is its holonym.

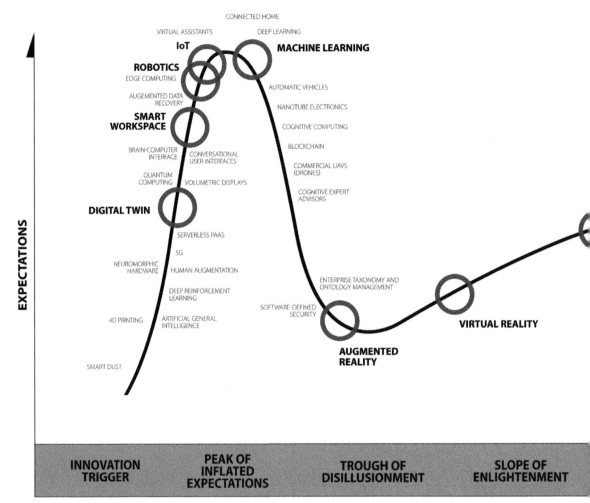

PROCESS MODELING AND TECHNOLOGICAL ONTOLOGY

Information modeling as it relates to building is a flexible, and expandable platform, and should not be understood solely as a new technological interface, such as computer-aided drafting (CAD) was in the 1980s and 1990s. While some of the design world is still in a backward-adopting position, that is, it is still transitioning from CAD or simpler design and documentation systems to more robust ones that allow for digital design and virtual construction, others correctly understand it as a projective and flexible set of technologies that allows architects to better understand and inform their intentions with more automated feedback to our work including digital twinning, production automation through robotics, and smart workspaces. These are seen by some as supplanting conventional BIM.

Figure 1.2 Gartner Hype Cycle for Emerging Technologies, after UNStudio, 2017: The Gartner Hype Cycle for Emerging Technologies positions BIM and Virtual Design and Construction as mature, if not technologies that are at their plateau of productivity. Other AI-influenced technologies such as machine learning and smart office environments are "hyped."

Still, digital design and construction makes more transparent and *efficient* that which once was a slow and cumbersome process – building – but architects should not simply settle on such efficiencies. We should expand on them to examine larger and more persistent issues in the world, from affordable housing and more democratized urban planning to climate change and the intelligent utilization of (natural) resources. Used in novel ways, tools should not simply solve localized problems but better understand and contextualize the broad issues we face as humans. In this sense, these augmented technologies can be quite disruptive and have more broad consequences for our future – a position that also expands the traditional scope of architects.

PROCESS GIVES WAY TO AUTOMATED TECHNIQUES AND WORKFLOWS

DIGITAL DESIGN & CONSTRUCTION

It's been noted that technology has allowed for an expansion of the architectural design process to be more specifically codified as technique, in the service of new digital or computation operations; and as workflow, in utilizing those computational techniques in the service of collaboration with others in building or fabrication. The term *procedural modeling* utilizes rule-based techniques, usually applied sequentially, to transform a set of geometric data or larger environment. It has been used to support generative design schemas.

Process has also been part of architecture, and specifically *making*, for much longer. In his book *Four Historical Definitions of Architecture*, Steven Parcell builds on four aspects of architectural production in seeking to define the architect and our work.[4] Here Parcell's elaboration of the concept of *techné* is useful. Noting that the ancient Greeks had no specific word for either architecture or art, he cites the development of the term *techné* as the "cumulative set of abilities that the Greeks has acquired during their development into a civilized culture." *Techné* was not a set of skills, such as those a twentieth-century worker would repeat indefinitely on an assembly line, but more a codified, yet linguistically unarticulated, means of production that incorporated both material aptitude and cultural memory. Those who utilized *techné* were therefore very aware of the participation of their work into the larger socio-cultural context or *environment*. Importantly, Parcell also writes of strategies for circumventing limits, ensuring that the objects of *techné* could be *new* and *advance society*, whether that knowledge involved the making of baskets, or jewelry, or buildings, and especially in ancient Greece encompassed occupations that could be both *manual* and *intellectual*; in other words, both *making* and *thinking*.

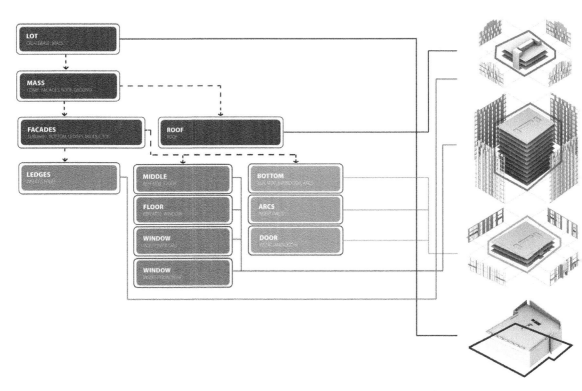

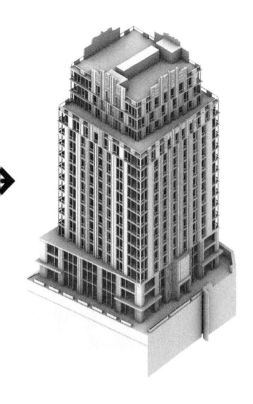

Figure 1.3 Procedural Modeling Basis, 2022: Procedural modeling is a method in generative three-dimensional design environments to create geometry based on a series of constrained and ordered steps. A series of recent academic papers have articulated procedural modeling schemas for the gaming industry where a "potential alternative to the labor-intensive modeling tasks required by traditional 3D modeling techniques for building reconstruction"[3] is advanced in the creation of a game environment. In the proposed method, the modeler starts with cartographic information and creates a series of subdivisions to a geometric mass to generate building features.

Figure 1.4 Mass-standardized assembly line, 1913: When Henry Ford introduced the assembly line in the early twentieth century, he revolutionized manufacturing so that products could be made more quickly and for significantly lower cost. This became associated with the dissolution of craft, as assembly line workers became exceedingly skilled at repeating the same, repetitive, operation without understanding an overall part-to-whole relationship, or interconnectedness, of the manufactured product.

TECHNOLOGICAL MODELS, AUTOMATED RANGES

Architects have been versed in some sort of model-making for the extent of our professional lives. We have used models to understand assumptions we have made about appropriate scale and form, put our models outside to ascertain shadow casting, and tested how they have fit into various contexts. The use of computers to provide iteration or variation in a set of solutions or criteria has been around for some 30 years, since the utilization of animation tools in architectural design by Greg Lynn and others. Critics were quick to point out that such exercises many times led to an arbitrary selection of solutions – how is the geometry yielded in, say, frame 32 of an animation sequence a better solution than frame 53? Such critiques of these exercises led to a proliferation of work in using the computer as a justification tool, and research papers that privileged vindication over more complex ideas about place and space making. These were a signal that there was a move away from a more intuitive method of using digital tools - away from ideas of metaphor or formal speculation and more specifically to notions of practicality. Such exercises seemed more consumed with making such digital explorations as a kind of pseudo-scientific proof, as opposed to exposing the computer as a radically new tool to speculate with. Nonetheless, the decade of the 2010s saw the computer moving toward a justification tool, both in terms of formal organization and material choices.[5]

Current resources adequately outline how information technology has aided architects and the broader AEC spectrum in the realization, or more precisely, actualization of complex buildings. Over past 20 years, BIM technologies – including CATIA, Tekla, and Revit – have become common (and in many ways more pedantic) ways architects execute buildings. In the intervening years, many research papers concerning parametric design endeavored to define computational methodologies for "understanding" space and situations.

CAPITAL FLOWS AND TECHNOLOGICAL WORKFLOWS

It is really the agency of the designer, not the tool, and the sympathy that the designer projects onto a specific situation or condition that leads to understanding – we are interested here in design defining a set of possibilities before proposing discreet solutions. Where the power of parameterization, a robust set of tools that allows for quick and precise iteration – usually of geometry – for a specific and constrained condition, should be of specific interest is in its simulation capacities. This can be loosely understood as a technological flow.

Figure 1.5 Mass-customized assembly line, 2022: As the assembly line as a production concept enters its second century, the unskilled laborers, versed at performing the same task repeatedly, have given way to automated robotic interfaces that can be programmed to repeat the same task ad infinitum or perform a sequence of bespoke operations. In this sense, we have left the twentieth-century paradigm of mass-standardization and entered a twenty-first-century one of mass-customization.

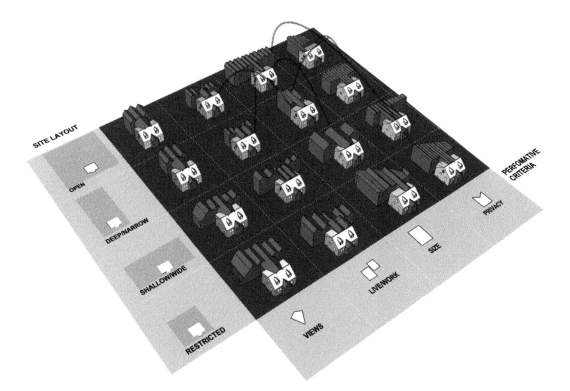

Figure 1.6 GRO Architects, Automated Range of Solutions for a Site, 2006: Animation tools first brought temporal change to formal solutions in the late 1990s. Since then, they have been replaced by scripting and graphic user interfaces that allow inputs to modify formal organizations in real time. Such workflows allowed architects to quickly study ranges of solutions, with variation being applied to a series of factors, including height, angle, and frequency over a given area.

Such a flow of information has been both exciting and transformative to the way we work, collaborate with others, and ultimately build; but these digital operations are only part of a larger and more complex equation that brings computing to the increasingly broad activities of designing and executing a building, such as site finding, formal explorations pertaining to zoning and client goals, and other "pre-engagement" services loosely categorized as "predesign"; as well as the type of information modeling that allows for such buildings to be financed – capital flows.

These flows of capital, in the form of currency, materials, and time, have traditionally been considered outside of the design scopes. Such information is increasingly being engaged by architects in design operations, utilizing temporal factors such as material procurement to inform design decisions more readily while also being able to more clearly articulate architectural features that are special or novel to a particular building proposal. Such actions begin to merge two types of information flows – those regarding *capital* and those regarding *aesthetics* – in ways architects have previously been unable to

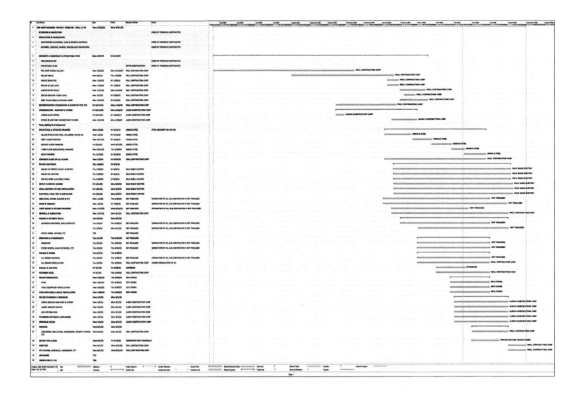

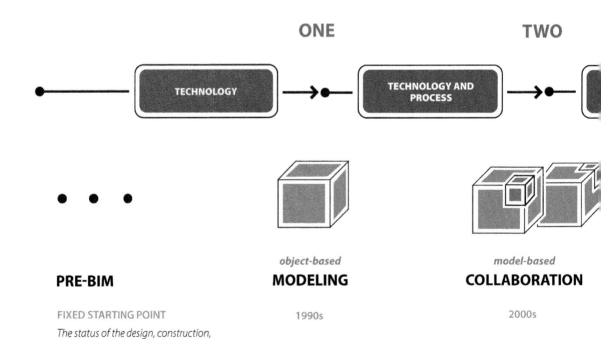

ONE

PRE-BIM

FIXED STARTING POINT

The status of the design, construction, and operation (DCO) industry prior to the proliferation of BIM concepts and tools; two-dimensional basis, interpretation abounds!

TECHNOLOGY

object-based
MODELING

1990s

TWO

TECHNOLOGY AND PROCESS

model-based
COLLABORATION

2000s

Figure 1.7 GRO Architects, Project Scheduling with BIM, 2021: BIM tools can connect to project management software such as Microsoft Project for temporal planning as well. This includes expansion into territories that are typically contractually relegated to the construction team. Metrics like a project schedule, organized in a Gantt chart, can be linked to three-dimensional models to track construction progress, anticipate material deliveries, inspections for financial draws, and coordinate trades.

access them. This is exciting not only in terms of expanding architectural territory or scope but also in how we can more successfully engage larger and more complex flows of data in useful ways. Specifically, how architects can access and merge their work with this expanded flow of information allows us to be both more comprehensive and precise in our work while positioning it relative to a more interconnected world.

Architectural design has always existed within an expanded field, borrowing from both art and other aesthetic practices, as well as the more science-based work of our engineer peers. This flow of information allows us to consider, and engage, an even larger heading of design computation, where both financial and aesthetic work is continually refined.

A SHIFT TO OBJECTIVE THINKING

Of late, the notion of this expanded field has facilitated a shift in the interest of some architects toward objects – as opposed to spaces but including *environments* – fueled in part by thinkers such as video game designer Ian Bogost and eco-critic Timothy Morton. This movement

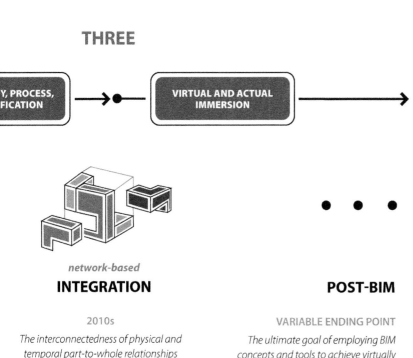

Figure 1.8 GRO Architects, BIM Capability and Assessment after BIMe Initiative, 2022: As BIM technologies matured, ideas about capability and collaboration abound. Early BIM schemas (2005–2012) focused on these within a model-based data set. Capability was largely focused on technology, or the soft- and hardware utilized in generating and delivering data; process, which sought to codify workflows as knowledge and skills leading to a product or project delivery and its management; and policy, which sought to develop a series of guides and set of best practices for establishing protocols within a BIM framework.[6] While such a platform did not explicitly exclude novel design operations linked to environmental simulation, resource consideration, or fabrication schemas, it also did not implicitly include them, making the assessment more a basis for software adoption within a conventional design-document tender-construction framework.

ON TECHNOLOGY | 14–15

does not privilege a humanist subject over non-human things that interact with humans. It is a rejection of the anthropocentrism that led to and carried most of the twentieth century. While some architecture-specific writing has been produced, the compatibility between such thinking and the digital tools and protocols being used by many architects has been mostly absent. Neil Leach attributes this interest in part to the fact that architects have always made objects[7]; however, a more holistic relationship to at least certain parts of this ontology can be made through a look into a more creative and exploratory use of these digital tools and protocols.

The evolving idea of typology in architecture lends support to such a shift to objective thinking. Leandro Madrazo argues that type is intrinsically related to the disembodied (epistemological) problem of form that stems from both Albertian questions of beauty and scientific questions raised about such aesthetics in the seventeenth century. In his doctoral thesis Madrazo writes, "The idea of type has much deeper implications than those that are confined to the classification and study of building forms. Type embraces transcendental issues of aesthetic, epistemological and metaphysical character; issues that have to do with the most generic problem of form."[8] Such ideas were carried through the twentieth century, when early digital theorists such as Greg Lynn began questioning the notion of typology in architecture through a rigorous study of geometry. Arguing that typology had been founded on fixed points based on harmony and order, Lynn proposed a more topological understanding of type that would be more nuanced, multiplicitous, pliant, and continuous – rather than single, fixed, and static.[9] Marcos and Swisher note this shift in their recent essay on notation; referring to Alberti, they suggest that part-to-whole relationships have been based on the (human) body, and if a body's fixed and individual parts are removed, a form of sympathy is broken.[10]

To adopt language from object-oriented programming, perhaps Lynn was arguing for an architectural typology that privileges an idea of *composition* over one of *inheritance* – the basing of a new object or type on the aggregation of its parts rather than a previous object. Such thinking naturally led Lynn, and many others at the time, to explore the multiple, and novel forms possible in three-dimensional modeling software. "Composition" in this sense should not be understood as a top-down arrangement of elements in building form but as the ingredients that make it up; thinking in part-to-whole relationships so that both designers and programmers can assign values to their work. In object-oriented programming (OOP), compositional-class objects can achieve polymorphism, allowing them to apply their values, or their pattern, across multiple types, privileging more flexible relationships and interactions as opposed to a final solution. While attempting to avoid strict analogies between Object-Oriented Programming (OOP) and Object-Oriented Philosophy (also OOP), it is ironic to note that the 1994 book *Design Patterns: Elements of Reusable Object-Oriented Software* refers to a program known as *The Builder*, which separates the "construction of a complex object from its representation so that the same construction process can create different representations."[11] The text even goes so far as to refer to Christopher Alexander's use of the term *pattern* on urban form and typology. "Even though Alexander was talking about patterns in buildings and towns, what he says is true about object-oriented design patterns. Our solutions are expressed in terms of objects and interfaces instead of walls and doors, but at the core of both kinds of patterns is a solution to a problem in a context."[12]

It is useful to consider a building information model as object-oriented in that objects making up the BIM (also an object itself) contain high-level definitions that refer to a database that structures the way it interacts with other objects, acknowledging an interconnectedness. In this sense, the object itself is not a static thing but something that can interact within and outside of itself. Objecthood, on the other hand, implies something stagnant – a model that is looked at as a kind of representation of something else. Structuring an active relationship with digital tools, and specifically their connectivity within the model and to external data flows allows the model itself to evolve over time. For Neil Leach, the tool presents "certain operations that it is incumbent on the user to recognize, depending on preexisting associations with that tool or object."[13] The development of the object in this case has as much to do with the finesse a designer can bring to the tool – how

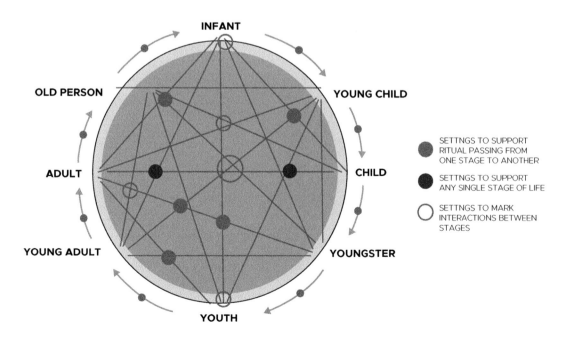

Figure 1.9 GRO Architects, The Full Cycle of Life, after Christopher Alexander, A Pattern Language, 1977, page 145. In his seminal work, Alexander and his team understood the interconnectedness of *things* within a network structure and offered organizations at various scales, including buildings and communities; and ways in which humans interacted with the "full slate of settings which best mark the ritual crossing of life from one stage to the next." Such passage is organized as a series of connected nodes.

the author interacts with the tool itself – as it does the interactions that the digital object will have with other objects. Leach continues:

> . . . there would appear to be little difference between 'tools' and 'digital tools.' They are all ultimately 'tools,' and instructions given to a construction laborer are not so dissimilar to code inputted in computational operations . . . An obvious example would be the progressive adoption of curvilinear forms within architectural design. With the advent of the computer, however, it became possible to define the curve very precisely.[14]

While the use of curvilinear geometry in architectural design is now certainly not new, the thought is not lost. Recall that a model already contains a certain *realism* based on its constructed relationships. An unexpected consequence of this in some instances has been that clients have commented that projects appear too "finished" early in the design process, a desirable condition in game design brought about through procedural modeling. It is possible that our clients expect to see representations and can therefore only understand the BIM model as one. The way BIM works can make things appear more real than they are in terms of a comprehensive design concept. A client used to having sketches presented to them to convey schematic ideas may very well be uncomfortable with the reality of a BIM, which in this case does not become more real, but becomes more precise as further associations and interactions are applied to it.

How a designer/author ultimately choses to structure the relationship between their work and the tools used in its service can be quite personal but needs due consideration. Whether this involves the plotting of points on a physical model to develop a virtual counterpart, as in digital twinning; the simulation of solar gain on a building, or the generation of machine code to produce a physical part; each require a specific relationship between parameters within the model itself and exterior parameters that act on it. Programmer Aden Evens makes a distinction that creativity is not solely an "expressive act" of a computer programmer but exists within the interactions between the coder and the computer itself.[15] The very nature of programming involves a not-specifically human-centric desire to create affects – allowing for spontaneous or otherwise unintended things (objects). It is critical that architects today "design" their relationship to tools. What the tools produce – either via a specific act or a generated or scripted one with an unintended outcome – needs to be both conceived and understood and suggests a difference between knowledge and data.

CONSIDERING TOOL-AUTHORING

The reader may draw the conclusion at this point that the designer today needs to be as versed in "software," as they are a creative aspect of their craft. This certainly has led to spirited exchanges between those who agree with such speculation and those who are ardently against it, arguing design remains a cerebral act that can be input by others following the lead of the designer – a common distinction made between the architects and draftsmen of the twentieth century. In fact, such spirited exchanges, unfortunately, remain in many US-schools where faculty many times fall on one side of this conversation.

Perhaps some of this discontent has to do with the "professional" nature of architectural practice, which was first codified by Alberti. In the US, to call oneself an architect requires going to an accredited school of architecture, obtaining either a bachelor's or master's degree, completing an internship, and then taking a licensure exam in the state where the architect will practice. Once licensed, licensure can in most instances be transferred to other states, as licensing requirements are somewhat standardized. While licensure is not required for everyone, many trained architects working at other's firms or teaching in schools need not be licensed; it is generally required to open a practice or submit drawings to an agency for the construction of a building. This has led to some overzealousness – I was once informed by a New Jersey architect that, despite being licensed elsewhere, I needed to be licensed in that state to call myself an architect there.

Writing about the licensure requirements of engineers, while calling attention to similar requirements of architects, Ian Bogost notes that "certification is usually offered only in fields where something could go terribly, horribly wrong with unqualified actors at the helm."[16] Bogost refers to architecture and engineering as civic professions as well as technical ones. Insofar as the use of tools, and specifically software, support this, he writes that it is not an "accident that the most truly engineered of software-engineering projects extend well beyond the computer," citing the construction of buildings, the design of vehicles, and the implementation of the roadway and energy systems that connect them together. The information-technology industry, however, not requiring nor valuing professional certification, has reframed engineering as business craftwork while neglecting the responsibility to the public that engineering entails. He accuses the technology industry of *engineerwashing* – labeling programmers as "engineers" to "make their products appear to engender trust, competence, and service in the public interest."

Bogost establishes a necessary link between creative work and the tools designers employ in its production. In this sense, and within reason, architects do have the opportunity to design their relationship to tools, or in the case of scripting, design the tools themselves. This, again, extends to formal generation and environmental simulation, where precise parameters form the relationship of a geometric assembly to a building site and its surroundings – potentially yielding new and unexpected outcomes in design work. This is one of the most radical changes in the way architects design buildings in the digital realm and redefines the nature of precision with respect to design speculations. Without such precision, we are left with Alberti's tradition, leaving only a human-centric way to interact with architecture. Indeed, in our current paradigm we have become more object-oriented, opening the possibilities for a more interconnected way in which we work.

Building information modeling can be seen as mundane if used in an "off-the-shelf" way. If a three-dimensional, data-driven environment simply replaces a more analog design workflow, without capitalizing on the power of such a platform, then it is fair to question the use of such digital tools. However, if the digital tools are used more creatively, furthering the architect's agency by engaging generative design potentials or engendering a virtual to actual fabrication schema (both of which involve some level of programming the digital tools) then the designer is operating in a wholly new way, establishing new protocols and even redefining the professional limits of what architects do. Though project scopes and schedules are somewhat out of our control, and may remain consistent with twentieth-century standards, information modeling allows us to change the way we work and how our projects are executed, allowing for more iteration and creativity along the way. While the most ardent supporters of a sort of conventional professionalism may disagree, we very well may be redefining what professionalism means to architecture in the twenty-first century while ensuring that creative authority remain a large part of the work of the architect.

There can seem to be disparate trends within the schema of modeling tools; however, these speak to a lack of conceptualization of the tools themselves and their use in the service of architectural design. As someone who has always been concerned with the relationship between my craft and the tools used in its service (by extension building), I am not quite sure which is worse, but these examples demonstrate an instrumental, and, again, ultimately consumerist, interest in a new technology – BIM – as opposed to the engagement of new digital tools in what should be a novel exploration and creative probing of design problems.

NOTES

1 Carlos Marcos and M. Swisher, "Measuring Knowledge, Notations, Words, Drawings, Projections, and Numbers," *Disegno* 7 (2020), pp. 31–42, p. 34.

2 Renate Motschnig-Pitrik and Jens Kaasboll, "Part-Whole Relationship Categories and Their Application in Object-Oriented Analysis," *IEEE Transactions on Knowledge and Data Engineering* 11 (5), (September/October 1999), pp. 779–797.

3 Gonzalo Besuievsky and Gustavo Patow, "Procedural Modeling Historical Buildings for Serious Games"; *Virtual Archeology Review* 4 (9), (November 2013); DOI 10.4995

4 Steven Parcell, *Four Historical Definitions of Architecture* (McGill-Queen's University Press, 2012).

5 Charles Eastman, Paul Teicholz, Rafael Sacks, and Kathleen Liston, *The BIM Handbook: A Guide to Building Information Modeling for Owners, Designers, Engineers, Contractors, and Facility Managers* (Hoboken, NJ: John Wiley & Sons, 2008), ISBN 978-1-119-28753-7.

6 See Capability Steps, by the BIMe Initiative, https://www.youtube.com/watch?v=NsSurYHNc7M referenced November 15, 2021.

7 Neil Leach, "Digital Tool Thinking: Object Oriented Ontology versus New Materialism," in *Posthuman Frontiers – Data, Designers, and Cognitive Machines* (Proceedings of the 36th Annual Conference of the Association for Computer Aided Design in Architecture), Ann Arbor, MI 2016, S. 344–51.

8 Leandro Madrazo, *The Concept of Type in Architecture – An Inquiry into the Nature of Architectural Form*; ETH Dissertation Nr. 11115, Zürich 1995.

9 Greg Lynn, "Deviant Types – From a Typological to a Topological Architecture," AA School of Architecture November 12, 1993, www.youtube.com/watch?v=Xc2eC43rlMQ accessed September 24, 2018.

10 Carlos Marcos and M. Swisher, "Measuring Knowledge, Notations, Words, Drawings, Projections, and Numbers," *Disegno* 7 (2020), pp. 31–42, p. 34.

11 Erich Gamma, Richard Helm, Ralph Johnson, Ralph E. Johnson, and John Vlissides, *Design Patterns: Elements of Reusable Object-Oriented Software*, ISBN 978-0201633610 (Addison-Wesley Professional, 1994), p. 97.

12 Ibid., page 15.

13 Neil Leach, "Digital Tool Thinking: Object Oriented Ontology versus New Materialism," in *Posthuman Frontiers – Data, Designers, and Cognitive Machines* (Proceedings of the 36th Annual Conference of the Association for Computer Aided Design in Architecture), Ann Arbor, MI, 2016, S. 344–51.
14 Ibid.
15 Aden Evens, "Object-Oriented Ontology, or Programming's Creative Fold," *Angelaki – Journal of the Theoretical Humanities* 1 (11) (2006), pp. 89–97, page 91.
16 Ian Bogost, "Programmers – Stop Calling Yourselves Engineers," *The Atlantic* (November 5, 2015), www.theatlantic.com/technology/archive/2015/11/programmersshould-not-call-themselvesengineers/414271/ accessed September 24, 2018.

IMAGES

pp 7–10, 13–15, 17 © GRO Architects; pp 11 © Omikron/Science Photo Library; pp 12 © RND Automation

2 | MORPHOSIS'S IMMATERIAL MOMENTS IN THE MAKING OF *THINGS*

Morphosis, Orange County Museum, Costa Mesa, CA, 2022: Two-dimensional drawing tender is still very much a part of the architectural design process at Morphosis, but is increasingly informed, and automated, by export and translation of three-dimensional data. In the firm's Orange County Museum, terra cotta panels were arrayed to follow the specific curvature of underlying façade geometry with minimal distortion. The façade is understood as a "design surface" which mediates between interior and exterior space, as well as site features. The surface is organized in both planar and one-dimensionally curving sections and relates back to surface geometry digitally generated or reconstructed in the CATIA environment.

Morphosis has a longstanding record of design excellence. Since 2005 the firm has been steadily augmenting its design process with digital tools that have allowed founder Thom Mayne to expand the firm's work in various building typologies and scales, as well as in terms of sophistication of design content and delivery, especially with respect to environmental and other requirements of institutional and civic clients.

Our story starts in 2012, with the recently completed Perot Museum of Nature and Science, in Dallas, and the team that performed all the BIM and parametric modeling for that project (see my 2014 book, *BIM Design: Realising the Creative Potential of Building Information Modeling*). Kerenza Harris, the firm's Director of Design Technology and an Associate Principal, came to Morphosis just prior to work specifically on the firm's Phare Tower in Paris, bringing experience in CATIA from Gehry Associates. At the time the firm was grappling with the complexity of information and its tracking through design development, finding there to be too large a time lag between concept design and the actual resolving of geometry and updating of what they then referred to as a *master model*. The firm sought to use computation to close the gap in a more direct manner. The team saw Dallas as one realization of the commitment to scripting and more advanced computation being implemented in the firm's design process. Importantly, this included a more seamless phase transition from design to construction, originally documented in *BIM Design*.

FLEXIBLE MODELS AND IMMERSIVE ENVIRONMENTS

Today, the firm continues to promote a design process that, on the surface, Harris acknowledges still seems to be somewhat chaotic while locating it within a computational and organizational basis that can promote new design workflows and expand architectural agency without relegating control. The firm is devoted to exploring both real-time and immersive situations to forward this process as it continues to execute larger and more complex work. This interest in real-time solutions has to do with the design, translation, and ultimately communication of concepts between tools and platforms. Geometry in this sense encompasses form, idea, data, drawing, and product, taking on a multitude of facets as it becomes "intelligent enough" to accommodate these things concurrently.

The firm's design process has always been understood as both progressive and comprehensive, but internally, there are transitional moments – between design phases, at drawing tender, commencing construction – that the computational team has sought to make more

Figure 2.1 Morphosis, Phare Tower, Courbevoie (Hauts-de-Seine), 2010: The irregular shape of the 300-meter-tall Phare Tower, or "Lighthouse Tower," was heavily scripted using three-dimensional modeling software, and responds to the shape of its asymmetrical site, which is defined by a neighboring roadway, metro station, and existing pedestrian walkway.

seamless. This has required the design teams to think about and build intelligence into their systems, as well as applying a kind of *global thinking* in terms of the systems themselves, especially in translating intent forward from early design phases; but also, during specific documentation phases and construction itself. This objective allowed design models to retain a flexibility as well as allowing them to respond to a variety of virtual inputs, from the internal design team as well as consultants and stakeholders, each of whom need to be communicated with in various ways.

In the past, design content would need to go through a "re-engineering" process each time specific data needed to be exported to a specific stakeholder from a model. More recently, Atsushi Sugiuchi works to ensure that design models have the interoperability required to effectively communicate all design intent with various external members of the team. Within the Morphosis workflow, certain trends or processes have emerged that are repetitive, where a direct connectivity to consultants has been achieved.

The immersive aspect of the firm's design workflow is more elusive. For the last several years, the firm has been actively engaged with a broader extended reality (XR) – encompassing virtual, augmented, and mixed reality – community, specifically with film makers and independent content creators. Design research has traditionally been applied during project design in focused ways; however, with extended reality they have found an opportunity to explore the *connectivity* between the design teams and the spaces they were creating, specifically how technology could assist in a more comprehensive self-analysis of their design proposals. This work led to questions of representation, and how specifically interactive environments could be represented to both the design team and users.

This interest in immersive environments largely commenced with the Orange County Museum and was also explored academically within courses being taught at the Southern California Institute of Architecture (Sci-ARC). Global real-time meshes that encoded various types and levels of project information have since become a goal of the firm. While this has been an ongoing process, with continuous improvements being made, the firm has applied this interest in the format of exhibition design, which is of a scale where collaborations with XR content creators could be expanded. In addition to a heightened experience of Morphosis projects, the firm sought to bring users into more close contact with specific design intent through an augmented virtual environment.

While the firm's "global design philosophy" remains unchanged and is still the purview of Mayne, Harris believes a more interactive basis for the design work allows both creators and users to understand and occupy spaces both physically and intellectually. This work has opened avenues of potential, and encompasses global issues where the power of more interactive modeling routines lies. In this space, a higher degree of information can be shared, processed and simulated for design decision-making, which is tied back to the various scales that Morphosis' projects engage. This allows both a local and more broad understanding of how a building functions and responds to the things it interacts with.

Such thinking has been applied to ecological concerns, such as terra cotta used as a façade system at the Orange County Museum to engage passively with climate and environment. The more the firm can refine its design process virtually, the more integrated data can be, and the broader its impact on architectural agency. Digital tools are allowing the firm to rethink design intent and relate that intent back to global philosophy. This increasingly includes allowing those who engage the firm's work as users to understand the buildings in more significant ways. Exhibition design has been a way to test assumptions about this process. To this end, the firm has collaborated with Kaleidoscope VR, a global promoter of independent, immersive content.

Renee Pinnell, the company's founder, has invited Morphosis to extended-reality events to speak about the consequences of such technologies on architecture. These events are largely made up of, and participated in by, gamers and filmmakers – not the regular, academic-minded architectural crowd the firm is used to engaging. The multidisciplinarity of these events permits interesting conversations about the aesthetics of space, its material (color, texture, density) attributes, and how it is understood by others.

The Morphosis team began to consider that emotional responses to space as imagined by architects may be completely different from those conceived by other content-creators, especially narrators and filmmakers – in experience as well as perception. The exchange of ideas that followed allowed the firm to work outside of the traditional scope and agency of architects, developing a series of exhibits.

Dassault Systèmes, the developer of CATIA, took notice of this emerging interest and invited the firm to contribute an exhibit for Milan Design Week in 2019, prior to the pandemic. The physically-based installation explored nonphysical phenomena within Morphosis's design process. Harris and her team created content and linked it to an app that allowed users to experience that content in a hybrid virtual/actual environment. The resultant work facilitated a strong shift in thinking about the firm's process.

EXPOSING WORKFLOWS IN MATERIAL CONSTRUCTS

As an architectural firm, Morphosis remains committed to making progressive buildings and physical environments. However, through XR, it now has ways to expose the largely hidden world of architectural and design-specific workflows. These sophisticated computational processes simulate and justify the design decisions that lead to a physical object – the building or environment. The organizational system behind this design thinking can now be revealed within the format of augmented reality. This capacity has ushered in self-analysis and raised questions regarding how much of a design process – a previously personal and many times esoteric exercise – is appropriate to expose through the hybrid reality of an augmented environment.

Within the practice, XR ushered in a more critical ability to consider its work that was beyond specific technologies, raising more basic questions about architectural *agency* and how the firm approaches *making* in the first place. The firm now actively participates in conferences and lectures that are outside of the specific disciplinarity of architecture.

Harris sees the integration of XR as a way of recontextualizing the firm's design process. Morphosis tends to develop their work based on part-to-whole organizational systems, such as an atrium as an architectural component within a larger assemblage, that occur in multiple projects in various ways. These systems allow for a familiarity given the firm's expanse of work dating back to the 1970s, as well as for engagement with new and complex conditions. This brings flexibility to the way design teams approach new work, being able to build on executed projects while having a sort of digital flexibility given the way these organizational models can process new information. It may be counterintuitive to expect augmented organizational control of a model or system to bring more flexibility and design freedom, but this is exactly what Harris believes the digital infrastructure does.

Control and flexibility have a long and intertwined history in architectural design, and for Morphosis, this *metaprocess* of activating and adopting underlying computational objects, which can be continuously differentiated and combined, produces new spatial conditions. The firm is developing a design platform rooted in the tradition of its own work, while allowing the freedom for that work to continuously develop. Here Atsushi Sugiuchi's computational work in interoperability is activated.

Sugiuchi primarily works in the development of software or routines that augment the flexibility of the firm's digital models, where this meta-procedural approach becomes apparent. Architecture has always appropriated tools from other industries and disciplines. We need only be reminded that BIM technologies were preceded by advances in geometric modeling tools borrowed from the film industry. It's logical that these modeling tools led that industry to augmented environments, which themselves have been adopted by the gaming industry. The merging of these worlds is not lost on Harris, who finds these interactions, which include those with film and video producer Alex McDowell, to be more beneficial in her work than the more mundane aspects of architectural specific software, which tend to privilege cloud-based collaboration. McDowell focuses on storytelling as a model that includes character development, recognizing that the portrayal of something that does not yet exist in the world is very much a way of enabling it to exist in the world – it encompasses a broad understanding of the world itself.[1] As a profession that largely consists of speculative or projective work, architects regularly oscillate between science fiction and science fact, already mediated by the virtual **real**-ness (not *reality*) of models. For Harris, this allows design strategies to become both broader and more direct – architects are no longer simply selecting tools but engaging more conceptually with narratives and user content, ensuring that the architecture of Morphosis tells stories that become actualized, or more precisely physical.

Specifically, Morphosis still seeks to utilize physio-spatial and nonphysical strategies in conveying design intent. In addition to more conventional programmatic functioning, the firm seeks to amplify users' social impact and experience. To this end, the firm is experimenting with sensory concepts that heighten a user's spatial awareness through lighting and sound. The goal is to make users aware of the choices they have. This might include taking a stair as opposed to an elevator to have a heightened sensory experience – again, the firm has utilized atria as "social mixers" in many of their projects. Morphosis projects have never been neutral, and the firm has a rich history of programmatic juxtapositions in its work. Yet Morphosis's work is an architecture of statements – social, political, and environmental positions – whose expression can be enhanced by nonphysical and sensory methods.

This move to immaterial means of expression is not in lieu of a material one. In fact, such attempts are ways of imparting more of a historical basis of each spatial experience within the context and culture of Morphosis's body of work. Spaces become more potent when contextualized, though the immaterial process also forms a kind of circular disruption in the genealogy of the work itself, allowing for both immediate and broad feedback when spaces can be understood both physically and within a cultural lineage of production.

Modeling – either physical or digital – as a design process is not simply an act of material aggregation but a three-dimensional process that allows the design team to think and visualize in three dimensions. It is as environmental as it is geometric. Design teams produce the *master model*, which is parsed and shared with others. Harris points out that this idea of nonphysical engagement is quite focused – key moments or spaces are chosen to be amplified in this manner, and these spaces require a higher degree of detail and representation.

It is these moments, perhaps 10 percent of a project's volume, where the teams deploy more advanced computational strategies and tools, and where the visualization and representation techniques take on nonphysical elements. This is achieved through components being activated through parametric links within the model that contain "behaviors," a kind of intelligence, that allow these components to react to criteria within the computational environment.

For instance, an atrium may take on awareness of external inputs and internal criteria. These objects are semi-autonomous and can adjust to produce information and respond to data that is fed to them. This might simply be a baseline change in geometric shape, much like a plant orienting toward the sun, or a more complex system that can be dialed to perform in specific ways. Designers allow these objects to "receive" information regarding climate or program so that the model can adjust. Over a building, these complex moments can build and begin to connect, literally weaving a spatial narrative.

As these systems become more complex, humans have more difficulty anticipating the results. Harris grants as designers they are setting up the experiments and have ideas about outcomes, so these are not unexpected; however, the process has yielded surprising effects and results. At some point, the iterations get to a level where something occurs that is beyond anticipation, something Mayne refers to as a "chance encounter." This idea of chance, a philosophical one, is sought by the design teams. It signals a point where the care and control of inputs within the model or system yields the unexpected. It is there where the buildings get interesting, where control yields to flexibility and newness – where the model is built to respond to a performance requirement yet generates the unexpected. The model meets the requirements input into the system, but the results both conform and point to a novel condition.

CHOICE AND CONSEQUENCE

Interestingly, this point of systemic development also reintroduces choice into the design, not the choice that went into the careful scripting of the system itself, a sort of meta-choice, but a kind of choice that may seem more traditional, relating back to aesthetics or affect after satisfying other criteria. It is a *both-and* moment for Morphosis. The process is a mix of a priori artistry and letting an algorithm do the work, trusting the technology to advance the design agenda.

The interoperability tools being built by the team and deployed within software environments, are primarily concerned with geometric manipulation. These tools are not software specific; they are not based within a single software package. This is where interoperability becomes important, as the team may, for instance, utilize a script within Rhino to produce a geometric affect that then is brought into CATIA for further development via hidden routines created by Sugiuchi. *Right tool for the task* remains a standard, and the ability to move between different software allows for a level of integration that fosters improved design development.

Harris and her team are always seeking a connection between the immersive reality research and an evolving yet individual design process that starts with pure geometry – digital geometry that exists without layers of metadata – but still has "query-able" information in terms of surface, edge, points, and translation data. While there have always existed file extensions and export types that allow for import and export into various software, functionality – the metadata associated with the geometry – does not translate. Extracting this information is part of the conversion process for Sugiuchi, who seeks to automate such processes so that teams do not have to manually convert hundreds of objects by layer. The idea that there are identical models within multiple platforms – itself an idea about the Digital Twin – is key to this work, and interoperability to facilitate updates to each when changes are made in one is a goal of this ongoing work in computation. In most instances, this work involves polysurfaces within the Rhinoceros, CATIA, and Revit interfaces, but also draws on some of the firm's earlier work that was developed in Bentley Systems' MicroStation. Design teams generally seek to translate their work to CATIA and Revit for more detailed development.

In this sense, the firm is redefining or perhaps augmenting architectural agency. From a geometric standpoint, and assuming agreement that virtual geometry is *materially real*, this process moves geometry from that state into an immaterial one to sort of revitalize it with augmented characteristics, to a new and steady material state through which it is tendered. Interoperability is important to understand procedurally, and through this actualization, when it is brought into the material world, it has the agency to produce both material and immaterial affects. This makes the geometry far more engaging to a user, even within this virtual state but particularly when it is actualized physically. Personal choice, something that perhaps occurs "by chance" within constructed buildings, can be studied within the model through highly customized, bespoke, applications that are flexible – they can be adapted to multiple instances within a system, similar to an automobile manufacturer's make and model platform.

To reinforce these goals, the firm has created, as an archive, an immersive environment of the firm's "dead projects." The catalog functions as a sort of repository of both ideas as well as virtual-material artifacts. While

Figure 2.2 Morphosis, Orange County Museum of Art, Site Plan, 2021: The museum is situated along the Avenue of the Arts in Costa Mesa and activates a large pedestrian area between it, other cultural buildings, and parking structures.

not standardized, it is possible for design teams to refer to projects and explore certain organizational concepts. Harris adds this "is not a just folder on a desktop" but access to real-time data, through a model, that agglomerates design content in a more seamless way.

ORANGE COUNTY MUSEUM

The new Orange County Museum of Art in Costa Mesa is programmed as a series of exhibition spaces with an educational component. The museum also includes a restaurant, café, museum shop, sculpture terrace, and back-of-house spaces. There is no parking within the structure – parking is handled through two nearby parking structures that seek to declutter the streets, making them more expressively urban in decidedly automobile-centric southern California. There is an important landscape component that serves as an urban gesture grounding a sculpture by Richard Serra, titled "Connector". Early design iterations sought to amplify the building's ultimate relationship with the artwork while proposing a multiblock linkage around the building to other cultural buildings along the Avenue of the Arts. Contrary to planning in many California downtown business districts, the building functions as a pedestrian hinge at the urban scale activating a large, multisite pedestrian plaza. The absence of automobiles creates the sense of a pedestrian density.

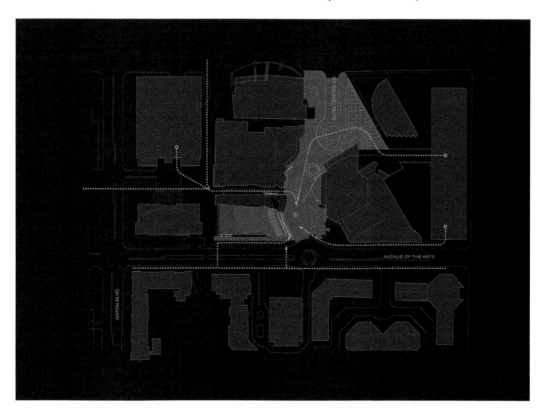

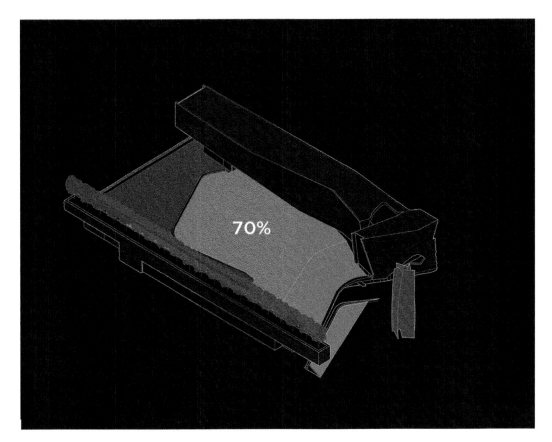

Figure 2.3 Morphosis, Orange County Museum of Art, Design Iterations, 2021: The building's front facade inflects around a vertical sculpture by Richard Serra, creating a raised landscaped promenade that is central to the building's transition from inside to out, and is referred to as a horizontal atrium.

The Serra sculpture functions to push the building, creating an implied enclosure on the plaza, so the building form bends around it. In addition to visitors sliding into this formal gesture, a stepped landscape exists above, where exterior functions are held. This open terrace covers almost 70 percent of the site, and functions almost like a horizontal atrium, as referenced in many of the firm's other buildings. The terrace is served by a monumental stair that, though not quite accessible from the ground plane, serves as a social space to animate the building itself, integrating with the architectural composition while generating an artificial green space, seemingly lifted from the context, which comprises 40 percent of the plinth.

While the museum program is relatively compact, the building's envelope is designed as an undulating and continuous ribbon, with an expanse of surface area that further emphasizes the building's relationship to the public spaces adjacent to and above it. Harris and her team became involved in the project in 2018, when the façade became both an urban gesture and a more generative aspect of the project. The ribbon's continuity helps link multiple aspects of the building and serves to connect the various scales the building engages while giving form to both interior and exterior spaces. While the design team liked the ribbon concept, it initially contained no texture or materiality, it was just a series of joined surface types.

Figure 2.4 Morphosis, Orange County Museum of Art, Design Iterations, 2012–2018: The museum project initially came to Morphosis in 2008. Over a period of about eight years, the building's formal organization changed significantly, while always in deference to the Serra sculpture at the building's entry. The building's envelope geometry was rigorously studied virtually and through 3D printing.

The design team, aware of the territory Morphosis has occupied, had to consider the complexity of the ribbon and the relationships it set up. Specifically, how that complexity could be expressed in the overall mass of the ribbon, rendering it as a monolithic façade element, or alternately expressed more subtly in the pattern and materiality that allows for the ribbon's resolution of curvature – the latter resulting in a more simplified formal language that is more locally articulated.

The ribbon started as a solid sculpting exercise using MicroStation, which through interoperability routines, a digital linkage was created to CATIA, where the surface condition was further developed. This script allowed the surfaces to be reconstructed within the CATIA environment as a series of intersected objects. Through surface analysis, points in space were defined for later construction development. Implicit in this step was the trimming of the rebuilt surfaces, while maintaining in the model the original, untrimmed geometry. This allowed the design team to "go back" in history to make local changes to the original surfaces while maintaining the trimmed relationship between connected edges. Harris considers these operations critical to anticipation in the project's development – what the geometry can become – its *potential* – is worked into the interoperability script in anticipation of further manipulation. This is an important component of the firm's digital design workflow and one that has

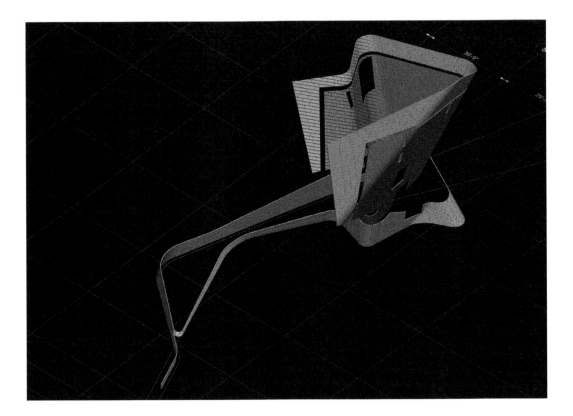

within it implicit consequences for the consideration of *making* – how something is ultimately built both digitally and physically.

Interoperability is an important factor in an expanded, and digital, architectural workflow today, and the ability to move seamlessly between soft- and hardware platforms not only streamlines a design-bid-build process but allows for the design and build team to work though problems of building in preconstruction (design) phases. In this sense, scripting as a design act today can be understood as similar to sketching in the 20th century and prior. It's at once a method for enhanced problem-solving and a clarifying action – making design intent more evident. It also occupies an augmented sense of design agency employed by the architect today.

This work becomes intricately connected to how members of the design team communicate their intent to others. Every detail produced to support design intent contains additional information related to the surface geometry. Implicit in this is the two- and three-dimensional location of the "design surface," which is the infinitely thin cartesian position of the surface, prior to offset or extrusion, as it relates to an overall material assembly at any location in the building. Pertinent intersections along the plane of the surface are isolated as work points and shared in communication of design intent.

Figure 2.5 Morphosis, Orange County Museum of Art, Design Iterations, 2021: Once the ribbon geometry was arrived at, the design team set out to bring both materiality and spatial linkages to it. The surface geometry was subdivided into a series of more manageable components, specifically joined planar and cylindrical surfaces, which were important to the ribbon's texture – its cladding materials. The geometry was also defined, and scheduled, by name and geometric type, giving the design team more useful information for more downstream construction operations.

Figure 2.6 Morphosis, Orange County Museum of Art, Design Iterations, 2021: Geometry was imported to the CATIA environment from Bentley MicroStation, where intersections between more primitive surface geometries were maintained. The design team was able to further manipulate the overall ribbon geometry through local modification of individual surfaces, which are then restitched through the interoperability script.

The team has found that if it released an exterior envelope model containing two basic sets of information – the location of the design surface and relevant work points and centerlines that pertain to it – it can track construction progress of the entire building façade. In this sense, the three-dimensional mesh of points becomes a kind of virtual infrastructure between the tender documents – the two- and three-dimensional drawings that dictate design intent to the construction team, and the construction work itself. Tangentially, the team can still test more local design ideas, such as joint patterning and material arrays, which were initially studied as a series of horizontal projections onto the ribbon surface itself.

This exercise presented a series of issues relating to the material resolution of the design surface. The complexity of the ribbon's curvature defied a horizontal material logic – while the one-dimensional curves could be projected onto the surface within the three-dimensional environment of CATIA, in practice it was materially impossible to do so with a horizontal solution. Projected lines, like the "design surface" itself utilized by the Morphosis team, are infinitely thin within the virtual environment and will ultimately receive some level of distortion through a material thickening operation, usually by extrusion or offset. This is an important aspect for consideration and is tied to the idea of issuing the design surface itself – the work points generated will always occur on an outside face of any given connection or will be dictated as offset from that point. Establishing such a practice brings a level of dimensional tolerance

that was not previously considered or achievable and demonstrates differences between conventional information conveyed in a document set and that possible within an augmented workflow, articulating issues of materiality, both practical and aesthetic.

Figure 2.7 Morphosis, Orange County Museum of Art, Design Iterations, 2021: Harris and her team initially studied a varied horizontal banding for the material resolution of the museum's envelope, but quickly found that the complexity of curvature within the overall surface would not allow such continuity. The team turned, literally, to other geometric solutions for tiling the surface, which included diagonal joints that would not distort (they could be kept constant) along the length of the surface.

Figure 2.8 Morphosis, Orange County Museum of Art, Intent and Constructability, 2022: Understanding terra cotta's physical properties led the Morphosis team to study a diagonal jointing, which was achieved by coordinating a projected pattern with planar and outward-sloping façade surfaces. By referring to the original cylindrical portions of the surface, a series of one-degree (straight) curves were located on the geometric surface that allows an overall panelization strategy.

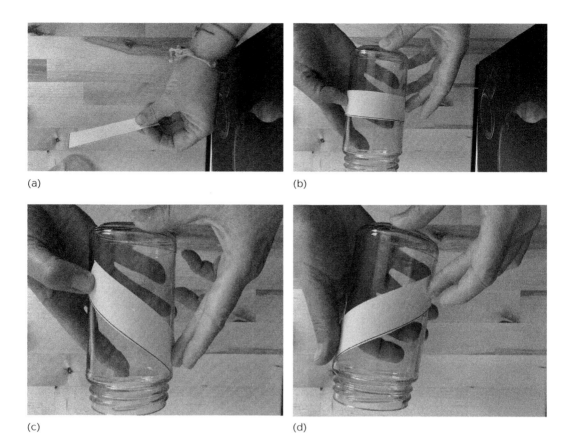

(a) (b) (c) (d)

Figure 2.9a, 9b, 9c, 9d Morphosis, Orange County Museum of Art, Design Iterations, 2021: A simple study in which a planar material – a sheet of paper – applied to a cylindrical stratum led the team to consider arraying the terra cotta diagonally across the façade as opposed to horizontally. Through coordination with the original digital surfaces, the team found that the diagonal array produced one-degree curvature – curving in a single direction – which is achievable with a cast cladding material such as terra cotta, as well as a piece of paper. This is the basic intelligence of the digital model that drives the pattern and its deployment on the geometry.

After several iterations, the team settled on terra cotta as the primary rain screen cladding system, and worked with an East Coast supplier, Boston Valley Terra Cotta, to better understand the material's capacities and behavior, and how it is installed. Initial work was conducted in a week-long workshop where the Morphosis team brought their façade concepts to the company. Materially, terra cotta is quite stable. It does not expand or contract significantly with temperature changes; however, because it is a natural material, production consistency is an issue. Morphosis ultimately settled on a diagonal joint pattern, which was studied by intersecting a desired grid pattern with the sloped and cylindrical geometry of the ribbon.

The team also considered tactile aspects of terra cotta, especially with respect to curving tiles, and they created a series of extruded dyes, fabricated from the digital model. The variation in the casting process was addressed through a constant temperature and humidity maintained in the production facility. Internal distortion in the panels could be technically controlled; however, the team found there was, in reality, some variability in the panels that occurred during the casting process. Again, the team referred to a distinction they made at the outset of the façade design process to have a local texturing

 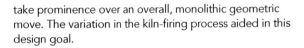 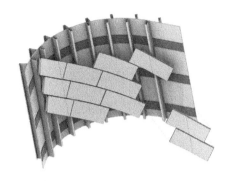

take prominence over an overall, monolithic geometric move. The variation in the kiln-firing process aided in this design goal.

Through this process, a digital script was developed that automated the forming and firing process of the material; however, the team still found tile distortion in the arraying of the material on the façade. Then came a subtle change – which changed everything. The diagonal laying of the tile on the façade surface allowed for single-directional curvature and largely removed distortion.

Referring to the original rebuilt CATIA geometry, the diagonal array was articulated at points where the surface changed direction, allowing the design team to resolve the curvature of the wall while minimizing distortion of the terra cotta. The aesthetic patterning also provided a kind of local movement, which broke down the overall envelope, allowing it to be read as a locally articulated material assemblage as opposed to an overall mass. This solution employed a "both/and" type of thinking as opposed to an "either/or" – the terra cotta was arrayed horizontally where the envelope was planar or mostly flat, while the diagonal assembly was utilized at point of cylindrical curvature.

Through the tiling process, the team found instances where repetition was possible. Ultimately, four primary panel types were utilized. The largest and most abundant is a 5'-0" × 1'-0" flat tile. Based on this proportion, three additional tiles, each 2'-6" × 1'-0" that are flat or partially curved, are utilized. Of these, two resolve the planar-to-cylindrical condition of the original surface geometry through a diagonally applied one-degree curvature. As seen in other projects, including the Perot Museum of Science and Nature, a simple rule-based system, and in this case with only four primary types, allows the firm to achieve variation over the scale of the entire building envelope.

Figure 2.10 Morphosis, Orange County Museum of Art, Design Iterations, 2021: A "both-and" strategy was arrived at to resolve the ribbon's tiling pattern that utilized the initial form-making captured in the digital model, imagined as a series of planar and cylindrical surfaces, and knowledge gained by working with the terra cotta supplier on the dye-making and fabrication process. The diagonal solution ultimately minimized the distortion and variation usually carried by a natural material. "Both-and" in this case also captures a partnership between the natural state of the material itself and a construction process that understands inherent material logic and applies it through a digital script to a façade solution. The solution is cost-effective, as it is not forcing the material to behave in ways it cannot, but it redefines the sense of "natural" through the human interaction within the design and fabrication process.

Figure 2.11a Morphosis, Orange County Museum of Art, Atrium Cladding, 2021: Atriums are regularly important architectural components within Morphosis buildings, with geometric and material resolutions augmenting these space's programmatic function as a social mixer. The undulating and sinuous form of the Orange County Museum, brought to the building's interior at the atrium shows the varied directionality of the terra cotta cladding system based on tiling rules on underlying geometry.

For Harris, allowing this rule-based system to populate the façade surfaces in a way relinquishes some control, referring to Mayne's desire to locate moments of "chance encounters" where the rules produce something that is unexpected. This is seen specifically in the building's vertical atrium, where the pattern is initially deployed horizontally, but the pattern resolution is ultimately driven by the shape of the underlying geometry and the curvature of its radii – controlled by the digital infrastructure put into place.

CONSTRUCTION

While the digital infrastructure deployed guides design development, it also plays an important role in the material production of the panel system. To facilitate more seamless communication between the architects and fabricators, Morphosis releases digital design data as well as traditionally printed two-dimensional tender sets, with specific guidelines to fabricators and subcontractors on

how to use digital data with the printed tender drawings. Here, *interoperability* again becomes important, not only given digital tender file types and compatibility but because the firm utilizes the capacities of multiple software packages to communicate construction information.

During construction, every surface that gets extracted from the original massing is unfolded. It is the unfolded state of the surface that is fitted within the tiling schedule. Information including cylinder rotation and radii, the

Figure 2.11b Morphosis, Orange County Museum of Art, Atrium Cladding, 2022: This diagonal resolution was translated from a virtual (rendered) to an actual (built) condition.

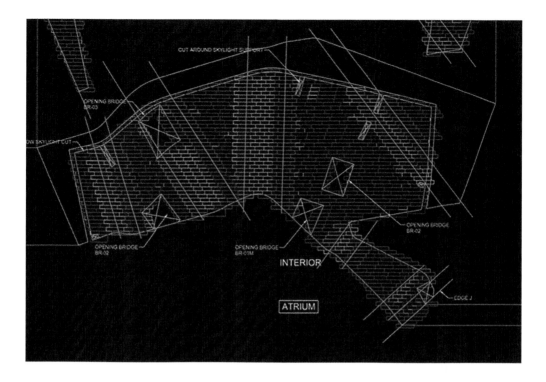

Figure 2.12 Morphosis, Orange County Museum of Art, Cladding Optimization, 2021: Optimization tools, written in both Rhinoceros and CATIA, guide panel adjustment based on tangency to cylindrical surface portions. These scripts allow the team to perform analysis of the cladding with respect to the underlying geometry, as well as provide a basis for the fabrication of the panels themselves through unfolding operations.

distance between tiles, and the return or connection to a planar surface is studied to finalize panel array and joint pattern. At this point in the process, all the surfaces are developable; the distortion seen in earlier digital iterations had been addressed so that a direct-to-fabrication schema can be undertaken. The pattern is aligned to the two-dimensional unfolded surfaces, with curvature located once it is rolled back into its three-dimensional configuration. While in some instances distortions still occurred, Morphosis's digital infrastructure ensured they were minimized, with greatest deformation from, or imperfection of, a surface being 5/16", well within a tolerance schema for building construction, but in some instances consequential for the design team.

Constraints were applied early in the process to wireframe geometry, with naming conventions and rotational values being the basis of further transformations. As each tile took on more specific metadata, they became either "surface" (planar) tiles or "volume" (cylindrical) tiles. These distinctions are critical for a streamlined production workflow. Drawing generation between Rhinoceros and CATIA begins with the automated export of geometric information between these packages, as well as their associated metrics.

Interoperability and the analysis it permits also become a sort of check of the design process and allows for a standardization of protocols across the firm, especially given the varied capacities of human designers and the software packages utilized. For Harris, this involves the "organization of language" that is malleable across both design teams and projects.

When the Morphosis team originally conceived of the façade, the infinitely thin design surface was understood as the front of the terra cotta panel surface. Through the fabrication process, it became clear that the design surface would be the *rear face* of the panel given the dimensional relation to the fastening substrate. The models needed to be adjusted based on this resolution.

Once produced, the tiles were fastened to a substructure consisting of steel clips attached to the primary structure with either light-gauge metal studs or furring channels. The primary structural steel support follows a "flat," gridded, logic. The tiles themselves received a corduroy finish, which reinforced the design team's initial concept of a horizontal ribbing for the envelope.

In addition to fabrication, BIM both checked and guided construction activities. Not only was the model used for more conventional checks such as clash detection, but on a more local level it was used to gauge deviation from the built condition in terms of the façade and secondary structure. Harris and her team isolated and tracked areas where deviations occurred and ensured they would be minimized per design intent. These areas resulted in some loss of legibility, or resolution, of the corduroy patterning in the panels. Since the design surface is based on the rear face of the terra cotta panels, tangency will shift slightly based on the geometric offset required to make the digital panel a solid. In the case of the façade, the tighter the radius – the more the panel curves over a smaller dimension – the more distortion a panel will receive in fabrication.

Just like in the original workshop attended with the terra cotta manufacturer, a series of more final mockups was created for the design team to study, with particular attention given to panels that curved cylindrically. The finish of the tile was a point of conversation, where the design team would in some instances clarify to the fabricator moments in the fabrication model. This speaks to the firm's agency and its ability to author complex production schemas to be undertaken by others. The museum opened in late-2022.

NOTE

1 Alex McDowell, "World Building," *Future of Story Telling* (2016), https://www.youtube.com/watch?v=0rwbWB8wsis, accessed 23 February 2022

IMAGES

pp 24, 38, 42, 45 © Morphosis Architects Credit: Jasmine Park; pp 26, 33–38, 39–41, 43 © Morphosis Architects

Figure 2.13 Morphosis, Orange County Museum of Art, Intent and Constructability, 2022: Design geometry is ultimately iterated from an original state that primarily serves to establish the overall envelope form to one of constructability. The development of a digital infrastructure ensures that the original design intent is maintained, while geometry gains intelligence. In the case of Orange County, all surfaces ultimately became developable, so a joint pattern could be applied two-dimensionally that would ultimately curve in a maximum of one direction when rolled back into its final three-dimensional configuration.

3 | ON TECHNOLOGY II

Twenty years ago, designers and scientists talked about simulations as though they faced a choice about using them. These days there is no pretense of choice. Theories are tested in simulation; the design of research laboratories takes shape around simulation and visualization technologies.
—Sherry Turkle, Simulation and Its Discontents, 2009

MECHANICAL SYMPATHY, CODING MATTER

In the 1970s, a term emerged within the elite Formula One auto-racing set. *Mechanical sympathy* suggested that drivers did not need to fully understand how their cars were assembled but needed to *live* with, from a performance point-of-view, the operation of their cars as an extension of themselves – that is, how they navigated as a *hybrid* high-performance yet human controlled *thing*. This notion of human extension, or post-human functionality, has existed for some time with the advent of prosthetics and similar advancements; but cars – and all other movement craft – have increasingly become automated sensing machines and do much of the work for us, including, in some instances, actually *driving*.

This is an amazing innovation, especially as ideas involving smart objects, including cities, involve sensing to monitor human actions. Sympathy has less to do with *control* than a willingness to cede a certain portion of it, while allowing us humans to still "design" a course – a Formula One driver needs to still navigate a raceway, which might mean cutting a corner or performing a dramatic pass, while allowing more conventional aspects of driving to be automated. In her book *Simulation and Its Discontents*, Sherry Turkle writes, ". . . in the future, creativity would not depend on understanding one's tools but on using them with finesse; the less one got tied up in the technical details of software, the freer one would be to focus on design."[1] To code or not to code, so to speak – how we navigate more advanced methods of executing our design work and employ digital tools to this end is an important aspect of twenty-first-century practice.

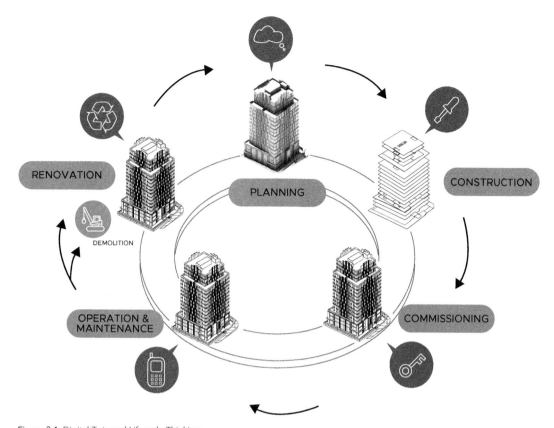

Figure 3.1 Digital Twin and Lifecycle Thinking, GRO Architects, 2022: The construction phase of a building remains a relatively small portion of a building's lifecycle, and the design process that leads to that even less still. The Digital Twin incorporates the proposed configuration of a building or environment, as well as its eventual as-built configuration, which would include any defects or deviations from original design intent. Digital Twins increasingly rely on digital data taken from real-time measurements, meaning they could be the basis of an expansion of architectural scope through construction into occupancy and post-first-use scenarios, including both demolition and adaptive reuse.

We should specifically consider information, and its relationship to hardware and physical processes, especially in the realm of construction or making. As BIM and its associated technologies further collapses the space between design and construction, Turkle's notion of *finesse* – or, specifically, the way we use tools as extensions of ourselves – is important, as are the sympathies they may impart. In living through a pandemic, we have further collapsed ways in which architects work via remote control through technology. That these processes and physical operations can be understood as extensions of our human-selves redefines the architect's *agency*.

Operations, which include the utilization of technology to interact more readily with engineering consultants and fabricators, establish specific costing or bills of materials, and explore more possibilities through iteration in our own work, certainly augment the work we as architects do, and consequently increase the rewards of our efforts. As the twentieth century closed, such operations became understood as a reassertion of the architect's control, where design professionals were reclaiming the space or territory that had been lost to others, such as construction managers, whose role was specifically imagined as engaging the messiness of manual twentieth-century construction processes. While this notion of *control* remains strong in terms of discrete professional scope, technology is indeed helping architects take on tasks that have conventionally been lost to others in the standardization of our profession in the twentieth century.

In considering the notion of our own agency and the critical opportunity to redefine it, the term *sympathy* resonates. This reclamation of territory indeed allows us to creatively reimagine ideas of work and production as they pertain to design and construction but raise important questions with regard to labor. Mechanical sympathy allows us to define tools in process, and especially the hardware, we use in making, which is increasingly a large- or full-scale operation. Gaston Blanchard has referred to this as the "objective mediation" occurring between researchers and their tools, which he suggested is how the researcher "makes progress" through new techniques and operations.[2]

OBJECT-PRECISION AND INFO-REALISM

The idea of precision as it relates to architectural design, and more importantly to the technologies we use in its service, is being redefined. While it is fair to call any architectural model (virtual or otherwise) "precise" – by its nature it has attributes such as dimension, volume, and degree of curvature that can be queried – the question must be raised about the necessity of precision, or its general level, or tolerance, that is required to execute a building. A virtual model is inherently precise, we can conventionally "measure lengths, areas, or volumes, although we can also measure weights, intensities, viscosities, etc."[3]

The more analog design methods that were codified by Alberti and others beginning in the fifteenth century and ultimately reached the height of their effectiveness in the late-twentieth century were contemplated in the service of an architectural subject – generally "man." These models were largely symbolic due to their limited (actual) precision and culminated in interpretation by architects, who over time increasingly worked at some distance from the sites of their buildings or projects. They increasingly in turn relied on others to build their work. Franklin Toker reminds us that by the fifteenth century, many architects were already working by remote control,[4] and the distance from the construction site, combined with twentieth-century construction means and methods, further removed our ability to ensure that design intent was properly met.

Virtual modeling has now eradicated many of the intrinsic limits of a noncomputational workflow. For one, thinkers like Manuel DeLanda and others have advocated that virtual models are already *real*, with material properties intact; they simply need to undergo some sort of actualization process to become physical. Next, and more importantly in the conceptualization of a computationally biased design process, virtual modeling is inherently object-oriented. It is both logical, though not necessarily linear, and performance-based; and it yields the creation of parts and assemblies that interact with other parts and assemblies to form a building, urban plan, or some other construction. Speculation on the actualization of these virtual objects has been topical to both manufacturing-bent designers as well as digital thinkers. Actualization in most instances has taken the form of digital fabrication, where computer hardware builds physical objects by a material process of addition, such as 3D printing, or subtraction, such as routing. These processes have found some sympathy with digital thinkers writing about objects.

Much has been written about models and specifically their use prior to design methods codified by Alberti, as well as their newfound, virtual, utility; however, it is useful here to take a more deliberate, and a more technical, turn to how software, and in particular virtual tools, are used to both generate and utilize models, and establish some concept of precision. In his essay "Brittleness and Bureaucracy," University of Warwick faculty Matt Spencer suggests two ways in which a "good" model can be described. The first case is rooted in a representational schema: a computer model that "generates data with a good level of fit with some comparison data set."[6] This way of

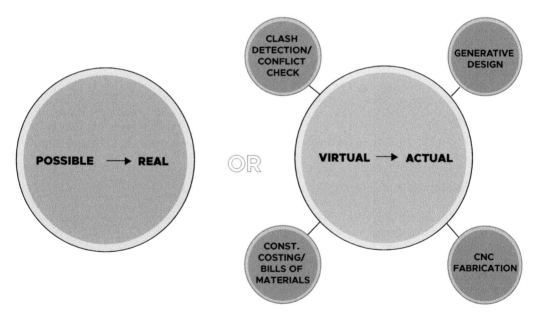

Figure 3.2 Virtual > Actual, Reimagined, GRO Architects, 2022: The Virtual > Actual was originally introduced in BIM Design (2014) as a way of capturing a more digital basis enabled by building information modeling. Levi Bryant has expanded the thinking of difference in the actualization process of a virtual object, suggesting that actualization cancels difference by stopping a virtual object, in effect canceling its range and its casual efficacy,[5] which has to do with its differentiation causing further variation within an environment. Actualization is a necessity of the architectural project and relates to the precision of the architectural object.

	IMPLICATIONS	EXPLICATIONS
VIRTUAL	Potent yet unactualized difference/Cause of beings/Pre-individual	Canceled difference/Formation of quality/Sterile being
ACTUAL	Condition/Cause of the actual	Product/Individual being without casual efficacy/Completion or end of process

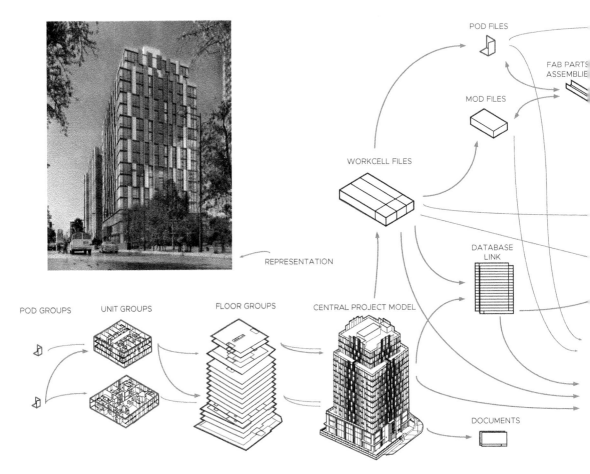

understanding a model as representation is contrasted with an appreciation of the model's basis for interactivity, its interconnectedness, and its "efficacious manipulative possibilities."[7]

Such a model can already be understood as sufficiently real, calling our attention to its properties, both material and effectual. The idea of the "real" emerged in the 1990s with the concept that virtual geometry (and by extension virtual buildings) is real – it has properties that can be queried such as mass and curvature. These models can interact with simulated environments. As such, they are always real but exist at different states of precision as they are developed. In this sense the model moves from producing a subjective representation to having objective properties that can be interacted with, by both a model-maker designer and other objects. Such an object-relationship schema has been codified by Luciano Floridi as *informational realism*, a view that the world is the totality of informational objects dynamically interacting with each other. Inherent in such models is the question of materiality, and how that materiality, when simulated performs and produces effects. Floridi uses the term *datum* to classify nonuniform portions of a model, where the model, for

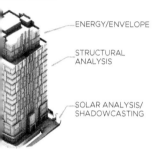

Figure 3.3 Jones *Kin*, Jersey City, NJ, GRO Architects, 2022: Matt Spencer, whose work sits at the intersection of anthropology, science and technology studies, has suggested that a "good model" conveys on one hand a level of representation, or "grounds for inferences about the thing represented," and on the other allows for optimizations in interactivity.

instance, changes curvature or undergoes some other material change. What is interesting in Floridi's version of object-oriented programming is the shift in focus from "the logic procedures, required to manipulate the objects to the objects that need to be manipulated"[8] themselves.

THE DIGITAL TWIN

The history of digital twin starts with NASA and its mission of space exploration. The notion was popularized initially as a simulation device, so that astronauts could effectively run a spacecraft, much like a flight simulator video game. Digital twins, however, quickly grew to be a virtual companion to a physical system, used not only to simulate function but to consider environmental and other factors that both the technology and humans using it needed to engage and understand. A digital twin is defined as a "comprehensive physical and functional description of a component, product or system, which includes more or less all information which could be useful in all – the current and subsequent – lifecycle phases."[9] Simulation was achieved initially via a physical model or simulacra but would ultimately come to merge the physical and virtual environments of the product though its lifecycle.

BIM systems today, with their massive capacity to organize data, are far more abstracted in that there are different container logics for different building components or systems – geometry, materials, and assembly means are all aspects of a design and construction process. While an argument can be made that larger information processing capacity allows for certain efficiencies – one has access to a great deal more information now than ever before – these seem to be exchanged for the directness, itself an efficiency, of prior two- to three-dimensional workflows.

This directness, a connectedness to geometric processes at scale, is very much back on the table today. The digital twin concept, which suggests less a one-way actualization process going from a virtual state to a physical one, speaks instead to a building more broadly "living" in two separate but interrelated states, a digital one and an actual one. This is a noted departure from ideas of material becoming, borne through translations via fabrication tools, where the actualization process involves the conversion of a digital model to the very code that allows it to undergo material formation processes to become physical parts. The concept has gained interest in manufacturing industries, and even the medical profession.

Figure 3.4 Apollo Simulators, Mission Control, Houston, 1969: Some 32 years before the term *digital twin* was officially recorded, NASA had a dedicated space for some 15 Apollo Simulators at Mission Control in Houston. These were exhaustively trained on by both astronauts and the mission control teams that would guide them through their missions.

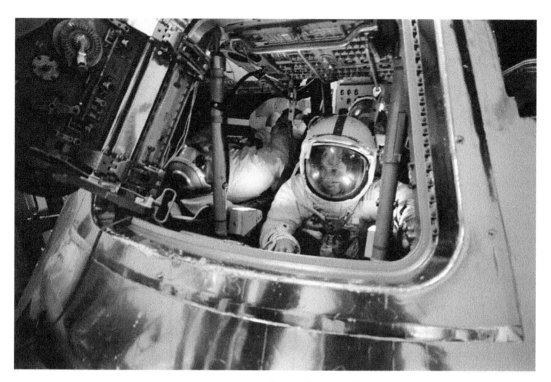

Figure 3.5 Apollo Command Module Mission Simulator, Mission Control, Houston, 1969: NASA provided physical simulators to mimic what they believed would be the operating environments astronauts would encounter on space missions. This included a kind of early flight simulator training. Ken Mattingly, who was bumped from the *Apollo 13* crew just days before liftoff on April 11, 1970, due to an exposure to measles, trained in the Apollo Command Module Mission Simulator. His exclusion from the mission would prove fortunate – he continued to work with the astronauts via the Simulator following an oxygen tank explosion that required emergency maneuvering to return the crew safely to Earth.

Within a robust digital twin environment, where parts and their assembly are represented at full scale, an immediacy of understanding is conveyed, but not every building information model is a digital twin. Models federated from a series of data to encompass design and construction scopes as well as operation logics can be understood as digital twins in architecture. The adoption of such a platform within architectural production is a biproduct of the Industry 4.0 movement (4IR), which supports increased automation of manufacturing and industrial processes and fosters increased communication capacities through smart technologies and the internet of things (IoT). Such a structure brings together teams of designers, engineers, and fabricators in an interconnected and global environment with loosely associated vendors who are activated based on market and supply chain logics. Digital technologies relate to this platform by allowing designers to expand architectural scope and practice while establishing new protocols of production and communication. This has allowed firms, such as UNStudio, to function as discreet architectural offices while collaborating with a global network of specialists.

Still a more dystopian view of such future practice, growing from the Ghost Kitchen trend, is that a large, malleable, and decentralized "practice," perhaps a group or groups of contractors operating in a distributed capacity begins monopolizing commissions within a typology or locale. Certainly, a decentralized practice is something most architects will be familiar with, given the pandemic years of 2020 through 2022. From a functional

Figure 3.6 SHoP Architects, New York, NY: SHoP's immersive co-VR space, known as *Tranquility*, is an interactive environment where multiple teams both within and remote to the physical space can collaborate simultaneously. The space was installed prior to the pandemic, and quickly became a favorite utility, especially as travel was limited in 2020 and 2021. *Tranquility* gets its name from Node 3, the module of the International Space Station that contains environmental control and life support systems.

standpoint, information technology has already moved architects to remote models of practice, either for data storage or shared environments.

A decentralized platform model enables remote team participation. However, it raises questions about the "human" aspects of collaboration, where personal connections are made. Firms such as SHoP have a fully functional co-virtual reality (VR) spaces, allowing remote design team members not only to collaborate but literally be in the same virtual space as avatars of themselves. Such a working method, from a technical standpoint, is a highly effective and efficient way of iterating conditions within a design model, especially when team members are remote from one another. John Cerone, a principal at SHoP, feels that these instances are most effective when collaborators can refer to a previous personal relationship, a sort of social efficiency that was forged in real space-time. Know your team member before you know their avatar, in a sense. Wolf Prix has similarly lamented in terms of his own preferred working methods throughout the pandemic but has come to enjoy the ability to sketch on live models via an interactive touch screen.[10] The technology fostering such collaboration existed before the pandemic and continues to become more seamless; but the behavioral aspects of such remote collaboration have yet to be fully considered, especially as a replacement for real-time in-person teamwork.

The digital twin concept has by now been architecturally tested on some notable projects, including those by SHoP and Morphosis, especially as it relates to modeling direct-to-fabrication schemas. As projects become larger and more complex, firms have begun to engage other ways to add value to building delivery. From an architecture standpoint, *value* is added through automation, more precision, and certainty; but how this value is actually calculated within a relationship to both owners and builders is less clear. The idea that everything will be produced through a high-level of CAD/CAM digital output is inevitable – it's already happening. Similar questions are being posed within the ecological space, with calls to consider environmental costs associated with building projects.[11]

The question is less about production and more about the model itself – that is, how can projects be set up in a more immersive model-based environment to allow for both easy collaboration and access to information for questions and decision-making? Cerone insists that this is more behavioral than technical.

To this end, architects are investing in both software development and real-time visualization to provide better access to design decisions within a more intuitive interface. This makes sense, as traditional methods of architectural output, including two-dimensional drawings and "dead" three-dimensional models, have a limited ability to convey intent to stakeholders residing outside of AEC disciplines. These participants aren't trained to "spin-around" within traditional three-dimensional environment; however, many are familiar with gaming interfaces.

This "gamification" of design content, especially within a digital twin environment, is a way to make the status of the design more accessible, and, by extension, to engage stakeholders more fully with the design process itself. The challenge with many gaming engines is that they are proprietary, so simultaneous integration still comes with challenges. By contrast, machine code, itself straightforward from a programming and generation standpoint, numerically represents *real* geometry, is a readily adopted standard by computer-numerically controlled (CNC) hardware, and allowed many popular modeling programs to accept plug-ins to output code that could be loaded to various CNC interfaces. Such an impetus is backed by manufacturing standards, themselves becoming increasingly automated, something that is not directly applicable to game engines created by and licensed through large software companies such as Microsoft and Sony. This presents a current challenge in the relationship between a game model and a manufacturing model.

SPECIFIC TOOLS FOR SPECIFIC OBJECTS

Architects have traditionally adopted tools for specific design tasks, favoring tools due to software functionality or personal preference, and utilizing a variety of digital tools in the execution of design intent. Software companies have been slower to allow interoperability without "lossy" file translation, though there has been some progress in this realm. Rhinoceros 3D is now more compatible with the Autodesk Revit environment through modules called *Rhino*.

Inside.Revit and *Grasshopper.Inside.Revit*, both developed by McNeel & Associates, to allow for "live" Rhino 3D components to be utilized within Revit in the modification and development of native Revit objects. Still, there remain challenges for augmented interfaces that seek to include stakeholders outside of design-specific scope, like clients or municipal review agents.

There is promise in interfaces such as UnReal Engine, Unity, and Nvidia's Omniverse in bringing about both a comprehensive *and* co-authored persistent and federated three-dimensional environment. The infrastructure of the digital interfaces focused on gaming, including remote user play, bring a more accessible virtual environment. These interfaces also introduce an alternative to NURBs modeling. Subdivision modeling, or SubD modeling, has already been ported to popular modeling packages such as Rhinoceros and Maya. SubD modeling seeks to minimize the use of of trimmed surfaces, which are expensive and prone to numerical error, to achieve smoothness.[12] SubD modeling allows the designer to manipulate a surface or solid by moving cage points, much like the manipulation of a NURBs curve.

Further, there is a "modular" kind of thinking at work here, as character rigs can already be translated into multiple game environments. But games need to preconfigure anticipated conditions and environments – they need their environments to be closed from a design point of view, leaving less to chance.

The digital twin, as a present object, has the ability to engage the traditionally projective nature of architects. It can at once *push the present forward* while *pulling the future backward* in this sense, as both an instructional tool for builders, and an assessment tool for owners and lenders. As such, the architect's agency is extended through the model's ability to "reverse engineer" a future physical (real) condition to test means and methods operations against economic and temporal efficiency. This suggests facility with a sort *of immaterial data flow* being a *future currency* of the architect. There are also broader implications for the continued involvement of the design team in post-occupancy scenarios, something many of the design professionals included here consider in their work.

As forward-thinking as AEC professionals seek to be, we collectively remain a conservative industry. The pandemic has been instructional in this regard in that it forced design teams to seek remote applications for problem solving. Still adaption to these conditions will be further challenged. In *Superusers*, author Randy Deutsch questions how this transition will be impacted by "cutthroat economic scenarios of the near future, including increased competition, rising automation, lowering wages despite increasing productivity, commoditized services, and thinning margins,"[13] ominous thinking for an industry comfortable with a "business as usual" approach to work. The digital twin is a viable way forward in that it codifies a series of design, construction, and use scenarios that allow the architect to be operationally present over the life of a building or environment, beyond the limited design scopes of the twentieth century, and further than the digital design and construction tropes witnessed at the beginning of the new millennium.

NOTES

1 Sherry Turkle, *Simulation and Its Discontents* (Cambridge: MIT Press, 2009), ISBN 9780262012706.
2 Gaston Bachelard, *The New Scientific Spirit* (Boston: Beacon Press, 1984), p. 171.
3 Carlos Marcos and Michael Swisher, "Measuring Knowledge, Notations, Words, Drawings, Projections, and Numbers," *Disegno* 7 (2020): pp. 31–42, p. 37.
4 Franklin Toker, "Gothic Architecture by Remote Control: An Illustrated Building Contract of 1340," *The Art Bulletin* 67 (1) (March 1985): pp. 67–95.
5 Levi Bryant, *The Democracy of Objects* (Open Humanities Press, 2011), p. 99.
6 Matt Spencer, "Brittleness and Bureaucracy: Software as a Material for Science," *Perspectives on Science* 23 (4), (2015): pp. 466–484, p. 467.
7 Ibid.
8 Floridi Luciano, Informational Realism, 2003. pp. 7–12. https://philarchive.org/rec/FLOIR.
9 S. Boschert and R. Rosen, "Digital Twin—The Simulation Aspect," in P. Hehenberger and D. Bradley (eds.), *Mechatronic Futures* (Springer, Cham, 2016), p. 61, https://doi.org/10.1007/978-3-319-32156-1_5.
10 Conversation with Wolf Prix, October 29, 2021.
11 Rebecca Mead, "Transforming Trees into Skyscrapers," *The New Yorker*, April 18, 2022.
12 Tony DeRose, Michael Kass, and Tien Truong, *Subdivision Surfaces in Character Animation* (Pixar Animation Studios, August 1998). Proceedings of SIGGRAPH 1998, https://www.cs.rpi.edu/~cutler/classes/advancedgraphics/S14/papers/derose_subdivision_98.pdf.
13 Randy Deutsch, *Superusers: Design Technology Specialists and the Future of Practice* (Routledge, 2019), p. xv, ISBN 9780815352600, 2019.

IMAGES

pp 47, 49–51 © GRO Architects; pp 52–53 © NASA; pp 54 © SHoP Architects

4 | UNSTUDIO'S FUTURE LIFECYCLES

UNStudio has utilized digital tools to maintain consistency in its work and constantly reinvests in technology to maintain the global reach of the firm. The firm's web presence lists on an initial splash that "we design with the future in mind."[1] This is not lost on Marc Hoppermann, the firm's former Head of Sustainability Engineering, who joined UNStudio in 2006. Hoppermann, who was also one of the Senior Architects for the firm's Booking.com project, notes that client expectations have changed following the pandemic of 2020–2022, especially as they relate to regular site visits. Since then, the world noticed that most of these interactions could occur in an online environment. With international travel mostly halted, the firm relied more heavily on video conferencing and other collaboration technologies. While this impacted the "rhythm" of the design teams in that interactions were now virtual and more scheduled than ad hoc, the firm noted that meetings became more efficient – team members logged in to discuss aspects of projects that needed to be reviewed, but time spent traveling or idling on project sites was minimized. As virtual meetings became routine, new habits formed, and many of the firm's members have come to enjoy the flexibility of partially working from home. While collaboration sometimes does present challenges – Hoppermann cites the frustration with not being able to sketch ideas while physically with other team members – the firm has remained productive within this hybridized production environment.

Technology, circularity, and modular building are integrated in the firm's recent work. In thinking about the future of building information modeling (BIM) and furthering the firm's interest in disruptive technologies, Hoppermann finds that the models become more central, where immense detail at a variety of scales is now possible. The vast information sets that the model now contains is almost unruly, so the firm has taken steps to ensure that intent can be conveyed to the contracting teams, this involves defining a language – codifying standards – that can be understood across the design and build team. Indeed, standardization and file and transfer protocols have been heavily debated by agencies such as NBIMS[3] in the US as well as software companies, which tend to have proprietary file types. As standards and protocols become further defined, architects can now "zoom out" again, considering the broader consequences of both the

Figure 4.1 UNStudio, Booking.com Headquarters, Amsterdam, NL, 2022: For all the intensity around the Internet of Things (IoT), and the more inclusive Web3, leading technology companies, in most cases, still need physical space. UNStudio has planned a significant mixed-use building project, a new urban campus in the center of Amsterdam that will be the headquarters for Booking.com, a Dutch technology company that "connects millions of travelers to memorable experiences, a variety of transportation options, and incredible places to stay".[2]

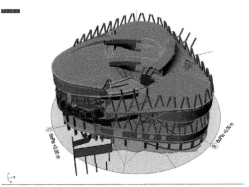
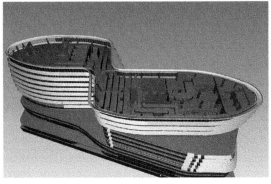

virtual model and actual building. In this sense, the model increasingly reflects the total lifecycle of a building, a kind of *meta-essence* of the work itself. For Hoppermann, lifecycle considerations are instantiated even before the virtual model is complete and conveyed, prior to the initiation of a project's construction. In earlier adoptions of BIM, the focus was on design, and specifically generative design capabilities, as well as translations to fabrication. Such a workflow allowed many novel buildings to be constructed with time, cost, and material efficiencies that would not have been otherwise possible. As modeling platforms have matured to contain even more information or data, the models can now exist within temporal scales that engage the time before project design, such as a predesign phase, and the time after building completion and occupation, a post-occupancy period. As most buildings are designed to far outlive the period of design and construction that bears them, and in many instances the original user of the building, models become important tools in considering building futures. UNStudio believes some of the most significant information that models are now able to convey address these predesign and postconstruction periods. For the firm, this expands smart building technologies or the digital twin analogy to encompass the building's lifecycle beyond a traditional use, and finance, period. In the Netherlands, a material passport is required to be tendered with all public buildings, allowing architects to ask questions about, and

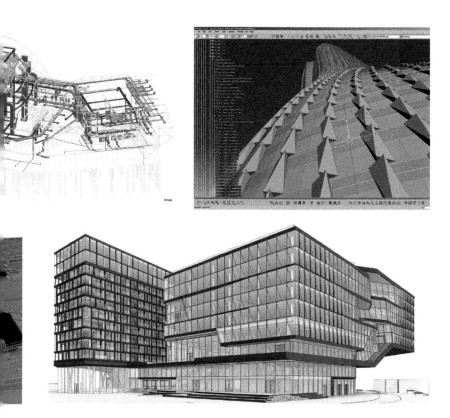

Figure 4.2 UNStudio, BIM Transition, 2009–18: Like many firms, the transition to a digital basis of design work occurred over time and was implemented based on need and experience with trial and error. The firm's work in this regard first started with the generative capacities of digital modeling, followed by integration with engineering trades and environmental simulation. It has evolved into a more precise thinking about building life cycles, and how architects can be involved in phases of building life following design and construction, concluding with the new campus for Booking.com.

understand, a building's material makeup and what can come of it in an adaptive reuse scenario. Hoppermann feels that artificial intelligence, robotics, and machine learning have only begun to integrate with BIM data and the architectural world in general.

For Hoppermann, a building's digital twin is not a solution in and of itself but allows architects to ask a series of questions of the building, encouraging ideas about the resources a building contains and those that can be added to these models to allow for new and future uses.

UNStudio co-founder Ben van Berkel very much sees new technologies in this regard less as readymade, but as concepts that deserve speculation in terms of their relationship to design.[4] UNStudio was early in bringing ideas about adaptability and flexibility into its design work, going back to its work on Arnhem Train Station,[5] where parking was combined with a bus and rail station, and commercial spaces. As the idea of flexibility at Arnhem was incredibly important to the project's sponsor, the firm sought new constructed methodologies to ensure a maximum flexibility in the successful distribution – and redistribution – of programs. The firm incorporated all the programs, as constructed moments, into the logic of a model. They eventually realized, however, that this sort of flexibility was a challenge to construction, so opted

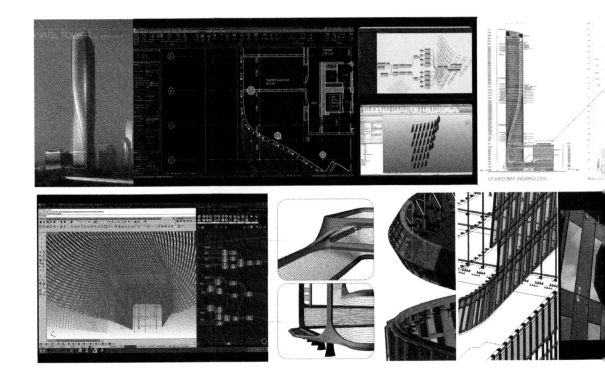

to utilize structural members that afforded the longest spans, allowing for a reconfigurability of walls and other architectural components within. This idea of maximum flexibility has been explored with their developer clients, who tend to want the most options for the longest period possible. There have been instances where the firm has maximized a building envelope only to be challenged by their clients to explore populating that envelope with different, and sometimes conflicting, uses responding to the uncertainty of the market.

These encounters have led UNStudio to explore various ways of anticipating such needs in their design work, finding ways in which technology can aid in ideas of flexibility and efficiency. Van Berkel suggests the firm has immensely benefited from these explorations, which have led to design innovation, not only in formal or organizational solutions pertaining to programming but also in the way consultants are engaged in the distribution of building infrastructure such as piping and ducting. Van Berkel saw as early as the late-1990s that design technologies would lead to new working models, with better collaboration methods. He has been particularly more interested in the nonlinear workflows models allow and amplify through new forms of collaboration. In borrowing from the late-French philosopher Bruno Latour, the firm sought to digitally understand capacities of various actors within a network, motivating a sort of hyperintelligence in a system.

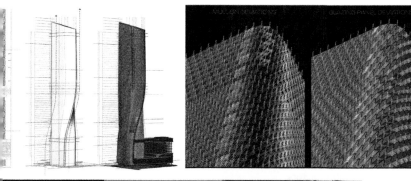

Figure 4.3 UNStudio, Al Wasl Tower, Dubai, UAE, 2018: Like any tool implemented within a critical and creative workflow, BIM was adopted gradually by UNStudio, only recently being fully implemented. The firm still relies heavily on the generative capacity of digital tools, especially for overall form and façade panelization, but are increasingly producing models that are utilized during design, construction and beyond.

Around this time, Latour, who had written broadly of both systems theory and ecology, outlined a series of conditions that he proposed would lead to hybrid networks of entirely new beings – hybrids of nature and culture. He called this condition a "Work of Translation."[6] He further suggested that cultures that have devoted themselves to the creation of hybrids have escaped the problems of being modern by understanding the differences between nature and (their) culture, a concept important to building futures.

With the advent of building information modeling some years later, a new technological language allowed for the codification of efficiency and flexibility, a sort of both/and in design and construction. Translation, after Latour, allowed for a continuous process of becoming that blurred the difference between the virtual model and the actualized object and sought to remove the interpretation (by others) involved in more analog approaches to building design realization. The technology validated the speculations undertaken by UNStudio in the 1990s and allowed the firm to create adaptive and interactive models on their notable early projects including Arnhem and the Mercedes-Benz Museum. Van Berkel notes that until then, architects in many instances were needed to "decorate" the exterior of a building. The speed with which UNStudio could engage questions posed by others on the design and construction team increased. Van Berkel reiterates that this enabled a new type of architectural control over both internal design criteria and organization, but also things generally seen as external to the digital model, such as cost constraints and material selection. By allowing others access to the model – drawing on his reading of Latour's networks – the architect was elevated to a critical actor in the design and construction process again. The impacts on design quality were immense, and fostered a motivational feedback to all actors invested in the model (project). Van Berkel is quick to point out that this is not a top-down approach or condition, but one that allows for various horizontal inputs. A reference to DeLanda's writing about genetic algorithms should be made here – that the architect sets up a series of variables that allow other actors a kind of read/write access, while providing a *meta-control* over the overall design and construction process.

Such a collaborative process also leads to what van Berkel refers to as "value-building" with their investor-clients. If the design team can convey how various uses and organizations are possible within a cost schema, they in turn create value over time and reduce the risk carried by the client. This intelligence comes with the model tendered by the architects – it is embedded in it, so utilizing the model as a flexible and adaptive construct *over the life of the building*, not just during design and construction phases, not only expands the scope of the architect but gives the building owner ideas about program, use, and even maintenance over time. Van Berkel sees such a practice extending the life of the building by a decade or more, which impacts investor profits as the initial investment is amortized over a longer period.

Requirements have been codified now in many institutional projects across Europe, where new concepts of control can emerge from these new demands – the intensified criteria owners put on architects. Energy use, or energy neutrality, and reduction of CO_2 emissions is something increasingly demanded in Germany and Holland, where van Berkel sees a direct relationship to circular thinking and sustainability. Health, specifically indoor heath, and lifestyle are equally considered, requiring a kind of localized customization of use that had not been possible.

Procedural modeling and related design and production technologies have further changed the way architects practice. Van Berkel welcomes a further integration of artificial intelligence with BIM, where circular use of the building can now be simulated. The models carry an educational capacity, as through them a user can track and better understand their own energy usage and habits. Such a technological basis has allowed architects to broaden our scope and collaborate with larger groups, which in some instances is outside of our core competencies. For instance, through simulation individual building users can see and understand graywater use and attempt to operate in an energy-neutral way. Across a community, this can impact a broader understanding of energy fitness, engender a better use of resources, and instill a certain amount responsibility to the community.

This has led to more recent ideas of connecting the building information model to building hardware ensuring that structures are both built and operated in an energy neutral manner. In their Brainport Smart District project in Helmond, documented in the *New York Times*, UNStudio rallies against the Smart City concept, which they felt was too abstract for a typical user.[7] The firm has been dedicated to improving social health through their projects which reinforce a kind of positive social dependency across a set of users.

Today, UNStudio is interested in allowing its design work to follow the lead of disruptive technologies, increasingly based on human–computer interfaces (HCI), allowing people to be more immersively connected with the spaces they occupy. While these technologies are not *form-giving*, they have broad implications to a user's experience and how that affects perceived qualities of space. Still, Hopperman sees some aspects of the broader architectural profession conventionally mated with the construction industry, and its unwillingness to adapt to change. By taking advantage of smart workplaces, automation, and machine-learning to facilitate design, and constructed, novelty, the firm is utilizing disruptive technologies in conjunction with more mature technologies like building information modeling and digital design.

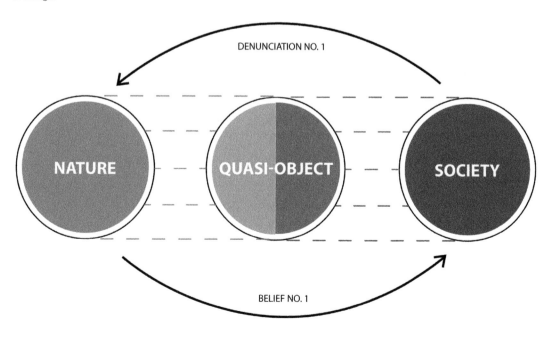

Figure 4.4 GRO Architects, Location of the Quasi-object, after Bruno Latour, 2022: In a diagram Latour referred to as the "locus of the quasi-object", he draws a distinction between the "hardness" of society and the "softness" of nature. He suggests that the quasi-object is arrived at through a hybridized network that emerges between the two extremes of nature and society. Latour writes, "if we have never been modern . . . the tortuous relations that we have maintained with . . . other nature-cultures would also be transformed." Architecture is positioned here, and the future begins to change.[8]

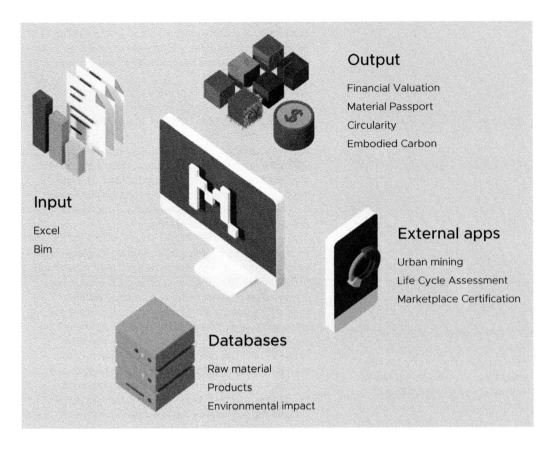

Figure 4.5 Madaster after BAMB, Digital Material Banks, 2022: Buildings as Material Banks (BAMB) is a European effort to utilize digital building data to optimize productivity, recycling, and residual material value to contribute to a more circular building economy. Through a database, material (physical) building data is mated with immaterial aspects of building life such as financing, servicing, and measuring of value.

In an expanded sense, UNStudio's use of building information modeling is a scalable platform, allowing its more general modeling and simulation capacities to be augmented with newer technologies. Still "BIM" for the firm also has a practical application as a tool used to instigate a set of tasks that are involved in the successful design and delivery of a building project. Such a practical position has allowed the firm to better understand, and design for, a full building lifecycle. Hoppermann is quick to remind that the construction period of a building is only a fraction of the time it will be in use, and the design and preconstruction period is an even smaller part of a building's life. How tools used in the service of design can be applied to an understanding of a building's life – its users, uses, and maintenance – has become of interest in the firm and is central to a larger post-occupation effort especially in Europe.

BIDIRECTIONAL LIFECYCLES

While the firm's earlier speculations in digital design remain important, and broadly impactful to our profession as they articulate a position on the relationship between digital design and making, the leveraging of data in consideration of longer building timescales has become central to the way the firm approaches new work. This lifecycle approach to a building, as opposed to focusing solely on the traditional scope of design delivery, is also consistent with a shift within Europe to understanding a building's internal and external relationships over a longer period.

For Hoppermann, "it's all data," which changes the way decisions are ultimately made. The component-based model can be analyzed for takeoffs and costing, more traditional BIM operations, as well as material and construction methods that specifically consider a component's resiliency and level of service or replacement. As such, the creation of the model, or database – expands the traditional scope of the architect. The model is used to understand how a building collectively ages, and anticipates changes in use or needs over time. Such thinking is consistent with the digital twinning occurring in other architectural practices, with bidirectional input from virtual model to actual building and vice versa. Within an environment encompassing virtual and actual states of a *real* building architects have the capacity to utilize data flows that transition between immaterial and material states to understand the building and its interactions in the world.

Such thinking radically differs from the paradigm of building delivery that existed through the twentieth century, when there was not only a primarily one-way flow of information from architect to builder, but also a series of *stops* at the end of each subsequent phase, whether design phases – a stop between schematic design and design development, at the completion of design documentation – a stop between drawing tender and the start of construction, and at construction conclusion itself. By making the entire process more seamless, UNStudio hopes not only to make better design decisions and understand more about their buildings, but in doing so increase the scope and presence of the architect beyond traditional design and document tender.

The firm's work has also seen a shift from more singular aspects of sustainable building components to more broad and impactful positions on ecology. Through models a deeper understanding of what a building can be and what it does allows architects to understand the building's positive and negative contributions to a broader built ecology.

BIM workflows are set up to understand components, their relationship to each other, their position within a construction sequence, and their residual value over time. In this sense, buildings can continue to be measured in terms of their environmental impact, and economic value can be ascribed to this fitness. Such a practice would bring together more disparate shareholders such as

Figure 4.6 UNStudio, Squint Opera, Unreal, BIG; SpaceForm, 2021: A collaboration between architects and gaming engine UnReal has produced a cloud-based Virtual Reality (VR) environment for design and collaboration. Especially given the pandemic of 2020–2022, real-time virtual collaboration spaces that allow for more than just design representation are allowing architects and design teams to further remote-control operations. Augmented reality (AR) within this environment allows for a perspective-, or walk-through, level understanding of building systems as they relate to spatial organization and finishes.

banks and financial institutions with design architects and engineers through the concept of a *material passport*:

> . . . a single passport would refer to a material, product or system. For a material it can define its value for recovery. For products and systems, it can define general characteristics that make them valuable for recovery such as their design for disassembly, but it can also describe specifics about a single product or system in its application. For instance, the connection of a product to a building is essential to understand its value for recovery.[9]

A broader thinking about materials gives way to new ways of thinking about material efficiency, as evidenced by the increased interest of architects in supply-chain logistics. Efficiency need not only be about initial cost or the logics of assembly, but how each figure into a larger understanding of schedule and, in the case of larger projects, impact on global resources. For instance, a material that is more readily accessible, and possibly therefore more economically viable for the then-circumstances of a building project, can change design thinking and direction. Hoppermann acknowledges such analysis is being applied to as-built conditions, but would obviously contribute to a more comprehensive understanding, and real-time lifecycle analysis, of a new building project. This idea impacts the circular economic understanding of material choices – one first needs to know what is inside a building before understanding what you can do with it. Such lifecyle thinking from the digital twin in that it is concerned with materials and use across an environment or system, as opposed to within a singular construction.

Hoppermann sees sensing as an extended reality that will ultimately link to and expand the utility of building information modeling in the firm's projects, especially post-construction when buildings are occupied. In recent projects, UNStudio is experimenting with smart building logics within their models. Linking sensors to the virtual models is a way to provide real-time feedback to the database, again an idea of bidirectional functionality. The architects and occupants can both understand and predict building performance. A simple example is how a building is responding to climate phenomena such as a heat wave. Occupants would access data to literally *see* airflow and understand costs associated with cooling, thereby making more informed, and economically tied, decisions to passive and active cooling schemas. Hoppermann also sees the opportunity for healthy competition within a community, with inhabitants seeing who is operating most efficiently within specific criteria, a concept echoed by van Berkel.

Hoppermann reminds us that UNStudio is committed to architecture – they are not a software company, or a company that produces sensing electronics, but UNSense, a sister company, tries to mediate between the spatial environments designed by UNStudio's architects and the electronics that can implement a higher level of experience within them. This is consistent with an expanded interest in user experience, certainly augmented through technology, that is being seen within the profession. More specifically UNSense is developing a data platform that can be implemented in the firm's buildings, where users opt in to information sharing that allows them, and various scales of a collective they interact with, to benefit from that data.

BOOKING.COM CAMPUS

Since 2015, UNStudio has been working with the city of Amsterdam and a private developer on the activation of a portion of reclaimed land adjacent to parcels given to the city's port activities. The site is directly east of Amsterdam Central Station, which serves about 250,000 transit riders daily. The bar-shaped tract of land has been the subject of a masterplan that seeks to correct the trajectory of planning that has, for Hoppermann, "turned its back" to the waterfront. The vision was to bring mixed-use programs – including commercial, retail, housing, as well as public functions – to the waterfront.

Booking.com, a Dutch technology company that connects users to destinations, first became interested in the parcel around that time, with a goal of consolidating

Figure 4.7 UNStudio, Booking.com Headquarters, Amsterdam, NL, 2022: A new campus for Booking.com in the heart of Amsterdam gave UNStudio the opportunity to explore its leverage of digital technologies in the delivery and potential lifecycle understanding of a building project while seeking to define future workplace conditions for a leading technology company. A sense of community and collaboration is experienced upon entering the campus via a horizontal communal space that sets up connections to various work programs. This large mixing space is populated with retail opportunities as well as an auditorium and conference center.

Urban massing

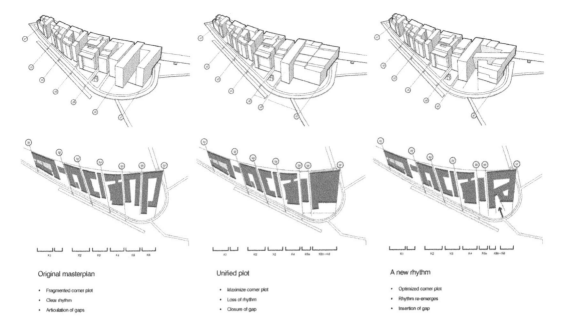

Original masterplan
- Fragmented corner plot
- Clear rhythm
- Articulation of gaps

Unified plot
- Maximize corner plot
- Loss of rhythm
- Closure of gap

A new rhythm
- Optimized corner plot
- Rhythm re-emerges
- Insertion of gap

its operations into a new urban campus in the center of the city. UNStudio originally performed a feasibility study for Booking, initiated by the landowner, and was ultimately awarded the contract for the new campus following a small, invited competition. The firm's work began with the engagement of the masterplan for the tract, which imagined a series of six lots subdivided based on footprint as well as a graduated density. The Booking site would be on one of the larger plots on a circular piece of land that terminates the reclaimed landmass at the River Ij.

In seeking to utilize the site for its new urban campus, Booking had three goals:

1. Improve the destinations where they are working and use their headquarters as a test for that goal.
2. Centralize operations on the site. Booking saw this as an act of community-building, and the creation of such a community was important to the way they hoped their workforce would utilize the campus – it was not to be a typical office building. .
3. Create a working environment that would allow for increased productivity.

The boundary conditions set forth by the municipality were rigid. Booking had strict program specifications, with about 100,000 square meters of office space for the new campus – a tall order for urban land, especially bound by water. While this number was ultimately reduced to 65,000 square meters, UNStudio's work focused on the subdivision of one of the eastern plots. Realizing that very

Figure 4.8 UNStudio, Booking.com Headquarters, Amsterdam, NL, 2022: Strict massing and subdivision requirements prescribed to the existing site required UNStudio to subdivide and then re-consolidate the site for the Booking.com urban campus. The firm created a structure that would contain enough floorplate area for the large amount of commercial space required by the technology company. This mass was ultimately eroded to make way for the vertical circulation schemas contained in the building.

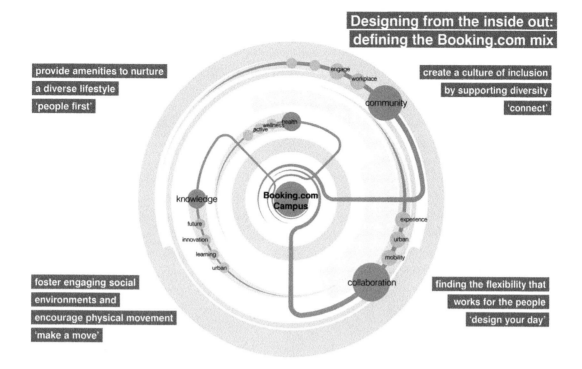

Figure 4.9 UNStudio, Booking.com Headquarters, Amsterdam, NL, 2022: UNStudio sought to bring ideas of knowledge, community, and collaboration to their proposal for the new urban campus headquarters for Booking.com in central Amsterdam. Fundamental to this is a "people first" idea that allows for diverse lifestyles and the fostering of both community and collaboration through a spatial experience within the building.

deep floorplates would be required to accommodate the large amount of commercial space, the firm ultimately settled on consolidating a portion of an adjacent lot with the end lot to make a parcel about 120 m × 120 m, which would be large enough to house the required program of the travel company. With this change, a 20 m cantilever was still needed to extend the upper floor plates to meet the area requirements set forth by the owner.

As the building form took shape, UNStudio devised a glazing strategy that would enforce the sculpted mass at the end of the reclaimed block while conveying the flexible and collaborative circulation strategy given to the building's interior through the placement of two atria. Specifically, while holding the overall glazing of the building consistent, variation was achieved through both mullion organization and the percentage of opening allowed in the façade, all within a prefabricated, unitized curtain wall system. Office spaces, which are connected to the internal circulation scheme, were treated with diagonal mullions that allowed for formal transitions between carved and planar façade volumes, while mullions at the building plinth, which set up the vertical thrust of circulation through the building, are treated vertically. In both instances, dropped ceilings transition diagonally up to horizontal mullions, allowing for maximum glazing height and optimized daylighting.

The housing component of the campus, which was ultimately reduced to 42 dwelling units, is varied further, with vertical apertures being framed with brick

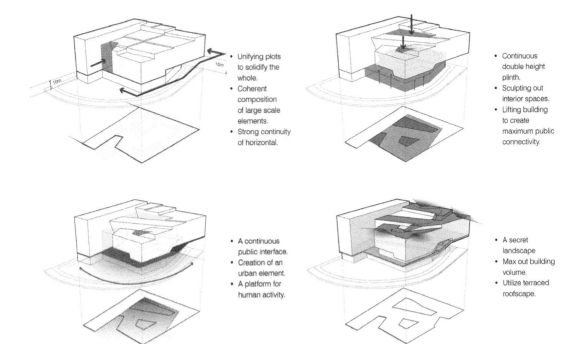

masonry, allowing for a different window patterning for the residential portion of the building. While Booking originally had an option to include the housing as part of its program, the developer opted to create upmarket residential units that are between 1,600 and 3,200 square feet (150 and 300 square meters) each.

CIRCULATION OF HUMANS AND SERVICES

The deep floorplates required presented other issues. UNStudio has typically utilized floorplates that were conducive to changes in use over time, so a commercial floorplate might ultimately be adapted for residential use. The size of floorplate required here not only precluded such considerations but challenged the design team to ensure adequate daylighting could be achieved. The resulting mass set the cantilevered office floors above a common double-height entry space and set up a public circulation route around and within the mass. The two connected atria were cut into the mass, the resultant shapes enforcing the public circulation route, which became part of the building's identity. The moves combined to give an upward thrust to the project, which contrasted its location at the terminus of the long horizontal band of buildings on the reclaimed tract of land.

The public circulation through the mass is programmed with amenities for the workforce, including an 800-seat cafeteria with views out to the city and over the water. Each atrium space is about 100 feet (30 m) tall but it enables pedestrian connections at each intermediate

Figures 4.10 and 4.11 UNStudio, Booking.com Headquarters, Amsterdam, NL, 2022: Once plots were consolidated, UNStudio sought to develop a continuous public interface that weaved diagonally through the resulting building mass. This circulation space began as a common entry within the double height plinth and then carried up through the larger mass which would hold the commercial floors. Two large and connected atria were cut into the mass that ultimately allowed for required daylighting while also extending exterior surface area to be programmed with office space, and providing views to the city and waterway.

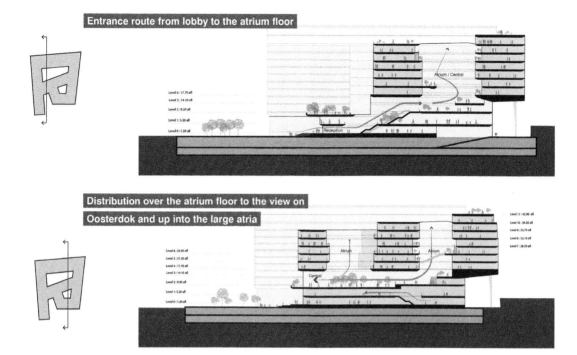

Figures 4.10 and 4.11(Continued)

Texture across different scales

1
Building volume

2
Volume is carved to create weaving image. This texturises the building at a large scale.

3
Facades are divided for panelisation.

4
Three-dimensional panels are placed around corners of buildings to create rippling effect emanating from the weaving areas. Besides addressing environmental factors, this texturises the building at a small scale.

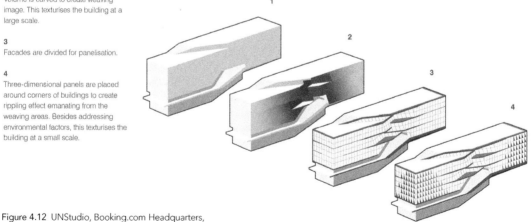

Figure 4.12 UNStudio, Booking.com Headquarters, Amsterdam, NL, 2022: Given office area requirements set forth by Booking, UNStudio arrived at a building massing with very deep floorplates. The design team set atria into the mass to break it up and increase daylighting opportunities. Additionally, façades were able to inflect to relate to internal circulation given to flex space that encompassed the office blocks. These facades, all unitized, were treated with different mullion strategies to enforce inflection and convey variation of internal organization on the exterior of the building.

 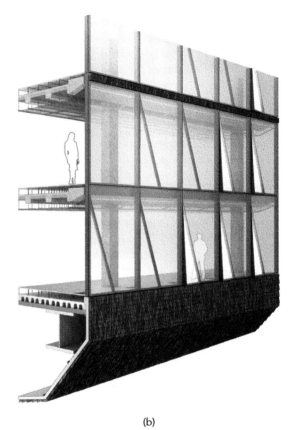

(a)   (b)

level. These divisions are planted and have abundant natural lighting, with the hopes of promoting community occupation and encouraging movement. Balconies and other amenities bring a human scale, breaking the would-be flow of office space behind a rigid curtain wall, which Hoppermann feels "creates possibilities" for collaboration.

At the building entry, the double-height plinth – a common area with retail, conference center, and auditorium – sets up circulation into the atria through a series of escalators and stairs. The space is designed for flexibility and serves as a multipurpose area where building users come together to both start their day prior to moving on to various work activities and conclude the day before leaving the building. Circulation is visually queued and begins with a grand architectural stair, a gesture of scale at the ground floor that divides into smaller strands located throughout the open spaces in the office mass. Abundantly planted and receiving natural

Figures 4.13 UNStudio, Booking.com Headquarters, Amsterdam, NL, 2022: Variations in the unitized glazing system allowed various and interrelated programs, such as offices and flex circulation space, to be read on the building's exterior. The insertion of two atria into the deep building mass allowed circulation spaces to be carved into the building's façade while guiding internal organization and increasing the surface area available for daylighting opportunities.

(c)

(d)

light, the atria function as a green lung for the project that is coordinated with the building's broader mechanical scope, which provides temperate air at a constant flow from each floor, and radiant heating and cooling from the ceiling. Breakout spaces on each floor were given to a series of interior architecture firms to complete, each coordinating their work with the larger core-and shell goals set forth by UNStudio.

The comfort strategy is tied to the goal of optimizing the work environment to increase productivity, itself tied to a personal experience of space. Working with Techniplan Adviseurs, a Dutch MEP consultant, UNStudio was able to achieve a comfort Class-A rating for the building, even though its façade is fully glazed.

The mechanical strategy also ties into larger-scale efforts possible in the Dutch climate to reclaim and reuse heat energy. The entire block takes advantage of an Aquifer Thermal Energy Storage (AETS) strategy,

which flushes waste heat to subsurface aquifers during the warmer months and is able to draw it back up into the building at cooler times. The strategy allows for passive heat storage, with the primary energy cost being heat pumps to move the heat energy to and from the aquifer.

Figure 4.14 UNStudio, Booking.com Headquarters, Amsterdam, NL, 2022: The atria are conceived of as a green lung for the building, organizing the Booking workforce communally at the base of the building and allowing multiple paths of circulation up and into the office mass. Pedestrian walkways and escalators diagonally connect various levels, encouraging a walkable space meant to decrease reliance on elevators.

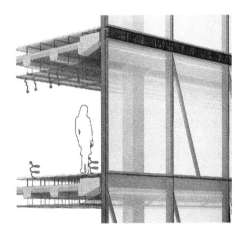

Figure 4.15 UNStudio, Booking.com Headquarters, Amsterdam, NL, 2022: The mechanical strategy developed by UNStudio and consulting firm Techniplan Adviseurs, was to provide constant temperate air thorough floor registers while delivering radiant heating and cooling to each floor through the ceiling plenum. The Booking campus was also tied to a larger heat storage and reclamation strategy for the block through underground thermal energy storage in subsurface aquifers.

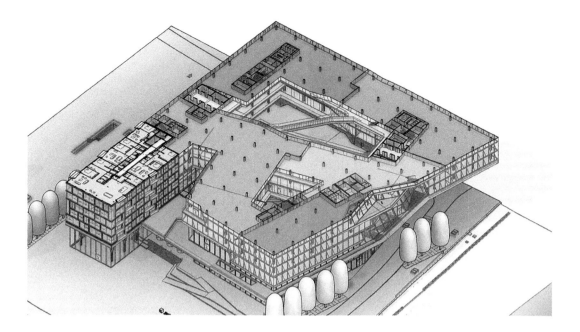

Figure 4.16 UNStudio, Booking.com Headquarters, Amsterdam, NL, 2022: Flexibility was built into the floor plates so that subdivisions for future tenants could be accomplished. The building's core and atrium logic allowed for different heating and cooling zones and the potential for eventual subdivision. The massing allows for a separation of the residential bar that permitted outdoor balconies where the office mass pulls away at a diagonal, visually splitting the housing and entry to the office programs.

LIFECYCLE CONCEPTS IN BOOKING

Hoppermann's interest in lifecycle thinking, especially as it ties into the utilization of BIM, matured through the five-year project development. Flexibility, again, was central to adaptive reuse ideas that went into the project; however, Hoppermann allows that the deepness of the floorplates – central to the Booking goal of consolidating its offices – complicates any discussion about the building's post-Booking future. Still, the design team sought to enable a multiuser capacity, subdividing the MEP system into a series of zones and allowing the building to have four to five discreet tenants per floor.

The project's brief also included considerations for a "BIM-ready lifecycle" as Booking's consultants created a database that established technical requirements for light, air, and comfort. Some of the data was directly linkable to UNStudio's model, allowing requirements to be digitally verifiable.

The central model, containing all structural, MEP, and building information, was shared within the Dalux platform. Dalux is a technology company based in Denmark that since 2005 has been delivering online digital tools that extend three-dimensional BIMs and their use in construction projects. While touting efficiency as one of the main goals for their work, they point out that efficiency minimized "time and resources wasted and lost, limiting pollution and emissions in the process."[10]

UNStudio ran a series of daylighting and circulation simulations in their model, which further guided design decisions and a local dialing of the mechanical system. Their model was ultimately uploaded to Dalux and utilized for project coordination, contracting assistance, and all

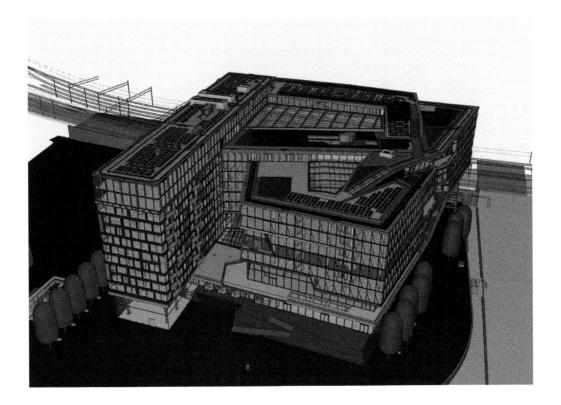

Figures 4.17 UNStudio, Booking.com Headquarters, Amsterdam, NL, 2022: Using the Dalux online platform, the UNStudio team was able to upload their building information model for enhanced field coordination during construction. The model allowed for plan overlays so user end goals could be understood while containing information utilized by building trades. The process allowed the firm to digitally meet multiple metrics set forth in a database provided by the owner, while allowing for plan-model coordination during project walkthroughs.

project walkthroughs. The Booking headquarters opened in late-2022.

NOTES

1 From https://www.unstudio.com/ accessed April 2, 2022.
2 From https://www.booking.com/content/about.html accessed January 18, 2022.
3 "Welcome to NBIMS-US," National Institute of Building Sciences, https://www.nationalbimstandard.org/
4 Conversation with Ben van Berkel, October 6, 2021.
5 Richard Garber, *No More Stopping* (Hoboken, NJ: John Wiley & Sons, 2016), https://www.wiley.com/en-us/Architecture+Timed%3A+Designing+with+Time+in+Mind-p-9781118910641.
6 Bruno Latour, *We Have Never Been Modern* (Cambridge, MA: Harvard University Press, 1991), pp. 11–12.
7 https://www.nytimes.com/2020/07/24/realestate/brainport-smart-district-takes-shape-in-the-netherlands.html.
8 Bruno Latour, *We Have Never Been Modern* (C. Porter, trans.) (Cambridge, MA: Harvard University Press, 1991), p. 11.
9 "Material Passports and Circular Economy," BAMB, https://www.bamb2020.eu/topics/materials-passports/circular/ accessed January 30, 2022.
10 From https://www.dalux.com/about/ accessed April 2, 2022.

IMAGES

pp 57–58, 78 © Klimaatservice Holland; pp 58–63, 68, 70–80 © UNStudio; pp 65 © GRO Architects; pp 66 © Madaster

PART 2

5 | ON ECOLOGY I

To be healthy is to be whole ... The task ahead is to elaborate the concept of wholeness in order to make us capable of overcoming the ideology of efficiency and prepare for a much healthier world, where we humans learn to make peace with the powers of Nature – in our minds and in our actions[1].
— Páll Skúlason, 2008

THE SCENE OF MAN

It is now common knowledge in architecture circles that the influential Dutch chemist Paul Crutzen claimed, in 2002, that we are no longer living in the Holocene, which is the climatic period that began after the last glaciations. Instead, he argued, with the steam engine, first practically invented in 1712 by Thomas Newcomen to pump floodwaters from coal mines, and later perfected for industry by James Watt with a separate condensing unit in the 1760s; we have entered into a wholly new and different epoch – the *Anthropocene* – which is characterized by man's industrialization of the world and a potentially dangerous rise in world carbon dioxide emissions. Such emissions have led to many environmental complications and degradations, including rising of sea levels and the loss of tundra permafrost due to rising temperatures. Shoreline areas, their low-lying landmasses, and areas built on reclaimed land have increasingly become subject to flooding and dramatic changes in topology. Shortly before Hurricane Katrina wreaked havoc on the southeastern United States in 2005, Elizabeth Kolbert noted this trend in the New Yorker, writing:

> In the seventeen-eighties, ice-core records show, carbon-dioxide levels stood at about two hundred and eighty parts per million. Give or take ten parts per million, this was the same level that they had been at two thousand years earlier, in the era of Julius Caesar, and two thousand years before that, at the time of Stonehenge, and two thousand years before that, at the founding of the first cities. When, subsequently, industrialization began to drive up CO_2 levels, they rose gradually at first—it took more than a hundred and fifty years to get to three hundred and fifteen parts per million—and then much more

Figure 5.1 Askja, Iceland, 2022: In a text about his visit to the volcano Askja the late Icelandic philosopher Páll Skúlason found himself criticizing efficiency, suggesting that the value ascribed to efficiency by our present civilization has become the ultimate value given the means at our disposal. Those means are the Earth, the whole Earth, and its various constituent parts including minerals, land, nonorganic lifeforms, cultures, collectives, and other resources that are finite in nature (literally). Skúlason saw inspiration in Askja, which defined for him human's relationship with nature.

rapidly. By the mid-nineteen-seventies, they had reached three hundred and thirty parts per million, and, by the mid-nineteen-nineties, three hundred and sixty parts per million. Just in the past decade, they have risen by as much—twenty parts per million—as they did during the previous ten thousand years of the Holocene.[2]

The late Icelandic philosopher Páll Skúlason uses the term *numinous*, which traditionally has had spiritual connotations, to make clear that the experience of nature, the numinous, does not happen discreetly in either our minds or in nature, but is brought into reality by our encounter with one another, under special circumstances.[3] To him, numinous qualities expose the superficiality of any system of ideas. He concludes that the discovery of nature is an objective exteriority of our subjective selves, as being(s) aware. Seeing nature involves understanding what we have done to it, what the extent of human habitation has meant for, and done to, the world.

VASTNESS

What does the inclusion of environmental or ecological aspects in the design of buildings allow for? Appropriately digital tools have allowed for the more precise simulation of environmental variables such as solar gain, shadow casting, and the effects of prevalent winds on buildings. Increased computing power allows simulations to go further, transitioning from building or building component simulation to those concerned with broader environmental variables. These working methods stand in stark contrast to those established by the humanist tradition, first codified by Alberti in his *De Re Aedificatoria* in 1452, which set forth protocols for architects and their work for hundreds of years. As documented in BIM Design, Alberti's treatise set forth human-centric design principals that also codified our work as distinct from construction practices. Buildings would be built by others who interpret the architect's representations, the design intent forming the basis of the construction process. Such a workflow saw resources as infinite, which was supported by the continuous rise of a consumerist culture in the twentieth century.

More recent work has pushed us beyond a subjective view of the world by calling attention to nonhuman phenomena such as global warming and other ecological crises that should fall under the consideration of architects. This objective form of thinking appreciates the interconnectedness of humans with other living things and the environments we inhabit, leading to what some refer to as built ecologies. We humans are also coming to realize for all its resiliency, Earth can be quite fragile.

Rebecca Mead notes that buildings are among the worst contributors to greenhouse gasses, citing evidence that 28 percent of global emissions are generated by building operations, and an additional 11 percent from the manufacturing of construction materials and from the construction process itself.[4] Much of this has to do with cement being used as the binding agent for reinforced concrete, which is the basis of virtually every new building foundation and most infrastructure projects. A recent study suggested that four billion tons of cement are produced annually. The chemical and thermal processes utilized in cement production are a major source of CO_2 emissions. It is also estimated that $90 trillion will be invested in infrastructure development through 2030, with two-thirds of that in developing countries. Moreover, the total global building floor area in 2016 was about 2.5 trillion square feet (235 billion square meters), and this amount is expected to double in the next 40 years.[5]

While this might sound like good news for architects and the broader AEC industry at a time when architects can rightly reassert ourselves into the broader territory of building construction, it is also becoming apparent that we are part of the problem. While the design phases of a construction project are a small percentage of the total time of construction, and total construction time is in turn a small percentage of the building's lifecycle, there is a disconnect here, and a longer view of efficiency must be contemplated. Perhaps the conventional efficiencies we bring to building, such as cost-effectiveness and novel means of construction, need to be expanded to create new propositions that relate to a building's larger, and longer, position in the world.

Figure 5.2 NOAA Sea Rise Viewer, 2023: A series of online simulators have been created since the 2010s to allow users to easily map the impact of a rise in coastal sea levels by intersecting a reference plane with topographic data. The website cautions that "today's flood will become tomorrow's high tide, as sea level rise will cause flooding to occur more frequently and last for longer durations of time." While much of the urban coastal zone of the greater New York region remains relatively protected at a 2'-0" rise in sea level, this volume of sea water will inundate lower-lying areas inland from the coast. While some of these areas already limit residential or similar uses, they contain industrial and infrastructural uses, with roads and rail lines connecting urban areas to the west.

While infrastructure is generally understood as roadways and bridges – land surfaces that connect us via automobile or other land-based transportation, reinforced concrete also finds use in coastal areas, which, according to a recent report from NASA[6], are also under increased threat. While there has been significant work done to combat storm surge, especially following a series of hurricanes that hit densely populated areas in the eastern and southern United States in the twenty-first century, NASA's Sea Level Change Science Team cautions that a larger concern will be the amplification of high-tide flooding due to sea level rise caused by climate change combined with the lunar cycle. The study suggests the accumulated effect over time will be more substantial, and costly, than a discreet storm surge event. The lunar cycle's duration is 18.6 years, with the Earth's tides suppressed over half of that period and amplified over the other. While the Moon's rotation and revolution has been steady throughout time, this consistency will affect rising sea levels. The impact will be the infiltration of flood waters inland.

Dense material systems designed for compression and land retention will still be needed at the bases of our structures and edges of our coasts – the physics will not change – but not thinking holistically about lifecycle and health of these structures at their time of design and actualization is a missed opportunity for architects. We have the opportunity to consider new material strategies as we engage climate change, and in doing so can expand our agency as we propose new built ecologies.

WHAT IS NATURAL ANYMORE?

The eco-critic Timothy Morton has developed the term *hyperobject* to describe phenomena that are so spatiotemporally distributed that they defy a literal understanding.[7] Hyperobjects, including weather events such as hurricanes or tidal flooding, only become partially recognizable during times of catastrophe. The computational power inherent in information modeling not only lets us better understand the data that drives these objects, such as temperature fluctuations in air or sea, or changes in wind speed or direction; information modeling allows us to visualize these hyperobjects in ways we were previously unable. Morton uses the term *futural* to describe this phenomenon; he further writes that hyperobjects are phased, that is, because of their massive spatio-temporal dimensionality, they appear and disappear from human perception over time, they are "impossible to see as a whole on a regular three-dimensional human-scale basis."[8] For instance, in the case of global warming – something architects are keenly aware of – Morton suggests that we only can see it in "snapshots," such as recent flooding in Houston in 2017 or the devastation in southwestern Florida particularly on Sanibel Island in 2022, both the result of two hurricanes, Harvey and Ian, respectively. Late summer or early fall hurricanes have become a regular occurrence on the eastern US mainland and meteorologists can produce simplified "plots" of these weather systems and their trajectories. A hurricane may be conceptually understood as an object within the hyperobject that is global warming. Using information modeling, we have been able to better visualize these objects, their effects, and how they interact with our work possibly limiting the anxiety that Morton suggests is a symptom of hyperobjects.

This is not to naively suggest that a building or plan can correct our current ecological crisis, but it is helpful to understand, and design, ways in which our future buildings will interact with these singularities. Such a platform would enable a kind of scalar thinking beyond individual sites to *vastness* – the scale to and at which humans have altered the planet. Morton suggests that the vastness of what he refers to as "geologic time" juxtaposed with the immediacy of our being is almost unresolvable; however, what if a building or plan could become more instrumental in managing environmental factors, albeit local ones. In this sense, designers could leverage the power of information modeling to visualize these things that were previously too large to understand, then simulate how they interact with our buildings and communities. This virtual playing

Figure 5.3 Astronaut training, Icelandic highlands, 1967: In preparing *Apollo 8* astronauts for the first planned mission to the moon, NASA used Earth – specifically, one of the few remaining environments on it that had not been subject to human activity – the high plains of the Icelandic north country – as a geologic training ground. The area was then one of the few areas on Earth that remained unchanged by human habitation patterns, and it was thought this would best prepare humans for extraterrestrial exploration.

out of these dynamic systems can allow us to better prepare our built environment to react to challenges by natural systems. Think of this as a computational extension to the rule-of-thumb guidelines advocated by environmentally conscious designers in the 1970s, but instead of assuming that one would, say, simply provide a perimeter drain around a structure in a flood zone; we can model how a building in the twenty-first century interacts with flood waters – for instance, how a building might purposely be designed to take on water to slow its recharge back into the ground. Such a design process allows for some level of innovation in the way designers propose new buildings and plans (objects) in the world.

Perhaps it is precisely these sorts of simulations that will further drive a new paradigm of architectural design. Performance modeling has already been a part of digital architectural workflows for the past 10 or 15 years, and has been utilized to achieve optimal solutions for a building itself, like how it will shed wind or minimize solar gain. The scale of these operations can be expanded beyond such local optimization schemas to allow an understanding of how buildings will react over time to environmental actions, and potentially allow them to mitigate these actions. What would such a building look like?

The homo-mensural principle, the idea that humankind is the measure of everything, established subjective truth – and perhaps even precluded an objective one. Measuring value solely against humans and our productivity impedes assigning values to resources and their loss. There have been attempts by economist Herman Daly and others within the last 50 years to quantify *environmental value*, which establishes cost based on a series of ecological factors that include the relationship between renewable and nonrenewable resources.[9] Higher resource prices would require more efficient ways of handling those resources, which could limit depletion and indirectly limit pollution. Such is an expansion of the importance of environmental cost beyond a conventional economic belief that an optimal allocation of resources is determined by the market. Some have even suggested depletion quotas on nonrenewable resources, such as petroleum, with pricing set on demand for a finite set of product. Quotas would generate cost based on a steady-state economic model that would put nonrenewable resources on par with, if not higher than, renewable substitutes; a model that is inconsistent with construction pricing – and pricing in general – within a Western capitalist model. For nonrenewables with "no close renewable substitute, the quota would reflect a purely *ethical* judgement concerning the relative important of present versus future wants."[10]

It seems such a position was an early form of nonhuman sympathy, and while boundaries are set within the market, those boundaries seek to reconcile efficiency with *equity* while achieving the "least sacrifice in terms of microlevel freedom and variability."[11] It should also be noted that quotas would guide an *appropriateness* of use, much like land-use maps and zoning control the limits of building. In most instances, owners of private land are free to develop their parcels as they wish, but within the confines of the state or municipality. The net-final cost in such a model is *throughput*, the flow of energy-matter from the natural ecosystem to the economy. Service, which Daley refers to as a *net psychic income*, is the net-final benefit.

Modeling and simulation technologies bring the opportunity to understand the interconnectedness of such resource management in an objective way. There are certainly aesthetic consequences to this shift. This object turn has been imagined by some as a post-natural condition. Maria Teresa Russo and Nicola Di Stefano argue that such a condition transforms the subjective nature of humans into an artefact. While there "may be an aesthetic of the artificial and of the artefact if it is in relation to objects, there is no aesthetic of the post-human body," and such non-bodies do not have a "harmony between matter and form"[13]; they are of a sort of post-human form.

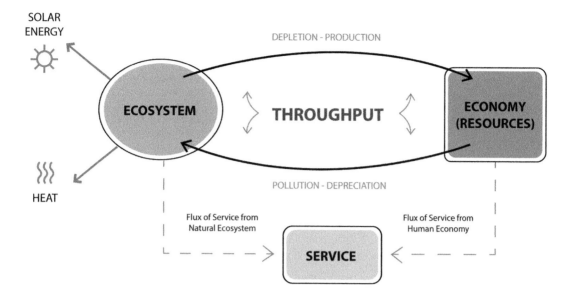

Figure 5.4 GRO Architects, Throughput in the Economy and Ecosystem, after Herman Daly, Steady-State Economics, 1991/2022: Throughput is defined in an economic model that assigns value to environmental costs as the flow of energy-matter from nature's low-entropy sources to the human economy and back to nature's high-entropy sinks.[12]

Many architects and designers came to the digital technologies that gave way to information modeling via the new formal possibilities it then-offered, so why not redress issues of form within this object turn? If we expect the new built environment to do things we have not asked of our buildings previously, then those buildings should be wholly new and different. With new requirements and criteria, buildings might become strange iterations of their former selves, no longer solely serving the subjects that inhabit them but confronting the objects they encounter in this post-human ecological space-world.

Sustaining implies that things can remain constantly *present* – in fact, another term for sustainability is *continuity*. Architects need to become more aware of the material consequences of our work – as others already have. Insurance companies know when our buildings will crumble; financial ones understand that they have an amortization rate. Some will extend beyond our lifespans.

THE ECOLOGICAL TWIN

In her recent book *Under a White Sky*, Elizabeth Kolbert writes about the challenges faced by a species of pupfish in the face of climate change.[14] The Devil's Hole pupfish is confined to a cavern of the same name in the Amargosa Valley, near Pahrump, Nevada, in what is now the Ash Meadows National Wildlife Refuge. Its entire ecosystem is about 8′ wide × 60′ long and connects through a maze of smaller caverns to an aquifer some 400 feet below the pool. It is believed that the Devil's Hole pupfish have been solely isolated to this habitat for some 10,000 to 20,000 years. As scientists became aware of this species and their highly specific habitat, they noticed ranges in

Figure 5.5 Devil's Hole is an approximately 8' wide × 60' long water-filled cavern, and an environment for a crucially endangered species of pupfish, *Cyprinodon diabolis*. The rock outcropping is filled with clear, warm water that connects to an underground aquifer and is geothermally stabilized at a temperature of 92.3 degrees.

Figure 5.6 Ash Meadows Fish Conservation Facility, Pahrump Valley, Nevada, 2013: The simulacra, or "fake" Devil's Hole, is a structure containing a 110,000-gallon pool of water which, given the anthropogenic change currently witnessed at the real Devil's Hole, is a lifeboat for the critically endangered pupfish.

the pupfish population, which correlated to water levels, which have been federally protected since 1976. The heat of the cavern remains at about 92.3 °F due to an overhang that allows a significant portion of its perimeter to remain in shadow at all times. The number of pupfish has varied drastically, but their number has never gotten that far above 100 fish. They have also dropped as low as 30, making them a highly endangered species. Multiple efforts to transport the fish to other "natural" habitats have been attempted in the last 20 years, but all failed, with no pupfish surviving.

The pupfish spawn on a rock shelf within the cavern that variably exists about a foot below the water surface. The platform is covered with algae, which the pupfish feed on. Scientists have found that the pupfish population correlates with the number of algae on the shelf, which in turn is affected by the amount of direct solar radiation received – Devil's Hole is primarily in shade, which helps maintain the constant water temperature in the Nevada desert. It's thought that anthropogenic-driven climate changes are causing a rise in the water temperature at the shallow shelf, allowing it to fluctuate to some 5 degrees higher than its usually consistent temperature.[15] This, in turn, is wreaking havoc on the pupfish's biological ability to spawn.

In 2013, the population of the pupfish dwindled to about 35, but by the time, through a $4.5 million grant, a fully operationally 1:1 scale simulacrum was under construction and opened that same year. The Ash Meadows Fish Conservation Facility is a structure containing a 110,000-gallon pool of water kept at the requisite 92.3 °F, with chemical and oxygen levels identical to those at the real Devil's Hole. This simulacrum is an ecological twin, containing a replica fiberglass-covered Styrofoam shelf for spawning. While it is only 22 feet deep, its perimeter size and contour resemble Devil's Hole, and a series of sensors drive a shading system to ensure that the semi-open louvered roof provides the same daily and seasonal solar exposure. Algae are grown in an adjacent facility and deposited on the shelf. The pupfish that live in "fake Devil's Hole" can live twice as long as their real environment counterparts, and they and the overall facility are continuously monitored though a series of cameras and sensors.

Reading about this immediately returned me to Jean Bauldrillard, who's work on simulation has become timely again. According to Bauldrillard, humans have replaced reality with simulacra, which in turn becomes a new reality.[16] Kolbert put it more succinctly: The Devil's Hole "simulacrum lies beyond the reach of human disruption because it's totally human."[17]

The Ash Meadows Fish Conservation Facility is novel not because it is a recreated natural environment – humans have been doing this for years. The idea of visiting a zoo or aquarium to see "animals in nature" merges two significant aspects of Western humanism in the twentieth century, the conquest of nature and a heightened awareness of leisure. The simulacrum doesn't allow visitors, but it does utilize technology to create a fully synthetic environment that, unlike its twin, is able to continue to thrive as an ecosystem. The simulacrum replaces the real, which in this case is an environment so constrained it only exists within a very small area on Earth. It very much becomes the reality for the pupfish thriving there, the encompassing synthetic nature that offers them a future path.

This assemblage meets the basic definition of architecture. It is an enclosing space providing a climate-controlled environment and it is synthetic, though constructed from mostly "natural" materials. In doing so, it conveys a sympathy for its inhabitants and their communal and environmental requirements. As such, the inhabitants can (somewhat) thrive in this architecture. Those inhabitants, though, are not humans, they are pupfish, and their survival is quite real.

Such work will increasingly fall under the scope of architects in the future, when we will not only be able to understand and engage environmental criteria for ourselves but also our nonhuman kin. In a way, we're all lost in the woods looking for nature, which is now mediated by the scene of man, even the pupfish.

NOTES

1 Páll Skúlason, "On the Spiritual Understanding of Nature," lecture given at Ohio Northern University, Ada, Ohio, April 15, 2008.
2 Elizabeth Kolbert, "The Climate of Man – III," *The New Yorker*, May 9, 2005.
3 Ibid.
4 Rebecca Mead, "Transforming Trees into Skyscrapers," *The New Yorker*, April 19, 2022.
5 Johanna Lehme and Felix Preston, "Making Concrete Change: Innovation in Low-Carbon Cement and Concrete," Chatham House, 2018. https://www.chathamhouse.org/2018/06/making-concrete-change-innovation-low-carbon-cement-and-concrete accessed 23 May 2022.
6 Carol Rasmussen, "Study Projects a Surge in Coastal Flooding, Starting in 2030s," July 7, 2021; https://www.nasa.gov/feature/jpl/study-projects-a-surge-in-coastal-flooding-starting-in-2030s accessed 24 May 2022.
7 Timothy Morton, Hyperobjects: Philosophy and Ecology after the End of the World (Minneapolis: University of Minnesota Press, 2013).
8 HyperObjects (Ibid.)
9 Herman Daly, *Steady-State Economics*, 2nd ed. (Washington, DC: Island Press, 1991); ISBN 978-1559630719.
10 Ibid., p. 65.
11 Ibid., p. 69.
12 Ibid., p. 35.
13 Nicole Di Stefano, et al., "Post-Human Body and Beauty," *Cuadernos de Bioética* XXV 2014/3.
14 Elizabeth Kolbert, *Under a White Sky: The Nature of the Future* (New York: Random House, 2022)
15 https://www.kcet.org/shows/artbound/mojave-project-devils-hole-simulacra, accessed 24 May 2022.
16 Jean Baudrillard, Simulacra and Simulation, Michigan, 1994
17 Kolbert, p. 84.

IMAGES

pp 83 © Visit North Iceland; pp 85 © NOAA Office for Coastal Management; pp 87 © Sverrir Pálsson, Apollo Geology Field Exercises in Iceland / The Exploration Museum; pp 89 © GRO Architects; pp 90 © Kim Stringfellow Projects 2022; pp 91 © Ash Meadows Fish Conservation Facility

6 | ZAHA HADID'S CIRCULAR ECONOMY

Figure 6.1 Roatán Próspera Residences, Zaha Hadid Architects, 2021: After years of enjoying success in the bespoke cultural building sector, Zaha Hadid Architects is experimenting with mass-customized modular construction solutions facilitated by a gaming engine to produce the Roatán Próspera Residences on Roatán Island, off the coast of Honduras.

Shajay Bhooshan finds himself speculating a little more on futures, with futures specifically being pluralized since the future can progress in any number of ways. In the work of Zaha Hadid Architects (ZHA), where Bhooshan is an Associate Director, he engages a digital design process that has within it a very specific, and creative, design intent that sees its outcome or actualization through a collaborative or co-authored "moment."[1]

At ZHA, there is a good deal of innovation that occurs on a project basis. The idea of collaborative moments or discoveries through interaction, especially as a research topic, is of particular interest to the firm. The nature of practice and the ethos of the company has always been *discursive*; they like testing and defending their ideas. The firm is actively looking to move from a traditional architect–subcontractor relationship to one of collaborative co-authoring that ties into the firm's idea of platform design, where they break the linear flow of a conventional design-tender process. The firm finds this traditional method to be siloed, which hampers the discovery process as discipline's interact less, and there is less opportunity for innovation. The longer arc of this research rests within architectural geometry, and specifically the relationship of geometry, through models, to construability and structural resistance.

DESIGN RESEARCH AT ZAHA HADID ARCHITECTS

Design research is a deeply ingrained focus of ZHA. The firm's 400-person staff, which Bhooshan is quick to point out is almost completely made up of architects and designers, functions as a creative brand, with each of the project teams being responsible for its own innovative approach to project design. Bhooshan notes that there are no in-house engineers or attorneys, effectively leading the architects to find solutions by raising questions, a practice that lends itself to design as a *research-based endeavor* as opposed to a *rote problem-solving process*. This sets up research opportunities as deeply collaborative interactions with outside consultants, allowing ideas to be nurtured further and making each of these efforts already co-authored by the time a solution is shared with the client. Through building actualization, these ideas are cataloged "back in the pipeline," allowing for a full-circle way of both understanding project-specific output and augmenting the techniques and tools the firm has at its disposal. Design research is generated by specific teams. Cross-team pollination also occurs, especially with the CODE group, where computationally generated research can permeate the activities of the individual design teams. This allows for a sort of *compositional feedback* to massing and formal organizations with Bhooshan's group structurally rationalizing many of the compositions generated by the design teams. The teams hold a good deal of tacit information, but over time this feeds into modeling methods and geometric approaches, as well as protocols for making drawings and diagrams. Successful projects ensure that both formal and representational schemes survive.

To this end, ZHA has an archive and exhibition team responsible for dissemination of information both internally and to the public. As a firm with up to 100 projects being developed simultaneously, this group is critical to sharing information across the design teams. An image and content library, organized as a searchable database has also been established, allowing the design teams to easily search for projects, sites, or geometric assemblages that share similarities to their current work. This evolving relationship to the work allows for ideas to vest and change over time – or be let go, registering criteria and constraints not only from singular projects but from the firm's evolving body of work, a feedback that is welcomed. A key role of more senior staff regarding this tacit body of information is to share with the younger design teams the firm's broad work catalog.

CONFIGURATION AND ASSEMBLY

Increasingly ZHA is seeking opportunities to apply complex geometric solutions, traditionally used in the firm's unique and bespoke cultural proposals, on a broader range of building types. Through small-scale efforts and demonstration projects, they have found that these ideas, combined with robotic manufacturing, can be scaled to have an impact on material savings and ideas about customization. Such work is not solely for the purposes of making unique shapes, but to solve or address problems of material consumption and customization from a social and user point of view. Buildings are therefore designed for specific uses, as opposed to being demolished or substantially reconfigured after a period because it is too generic – not customized to a specific user or use. Architectural geometry, with its "pre-computer" history can be repackaged into a platform design idea where customized, yet mass-manufactured floor slabs, columns, or other building components are possible.

ZHA has had a goal to engage large manufacturing companies in the creation of prefabricated components, in effect creating customizable products. Many of these manufacturing concerns, however, are less agile due to development and investment in specific pipelines or supply chains. To prove the concept, ZHA has taken to small pilot projects including a 13-unit housing development on a small island off the coast of Honduras. In this remote context, the client will not invest in technology unless it solves a specific problem. Component solutions must either save on material costs through a reduced section or support a local workforce or economy.

These solutions do provide a novel aesthetic, which ZHA is keenly aware does have value in the market. For the project, ZHA developed two pieces of technology specifically supporting a manufacturing process that allows for customization to occur at two very different scales.

On the backend is the concept of a *circular factory*, for which ZHA functions in an advisory role, on the island. As both commitment and investment in the project and locale, the owner imports robotic manufacturing capabilities and trains the local workforce to produce components from locally sourced materials – specifically timber. Trade skills, with materials such as concrete, necessary for building foundations, are also part of this process, making the factory a kind of hybrid machine of traditional knowledge and the application of advanced technologies that require a range of analog and virtual human interaction.

On the front end, ZHA developed what they dubbed a *configurator*, which incorporates all their prior coding knowledge and experience with architectural geometry into a game engine, so that user interaction is more natural, and less "CAD-like." Users select options, such as paying for air- or view-rights that allow for a customized and unique solution that contributes to exterior form, while also selecting from a palate of interior configurations and material finishes. The solution rests somewhere between the standardization principles that guided manufacturing in the twentieth century, and the mass-customization goals of the twenty-first. In the configurator, there is almost a meta-standardization that guides the organization of the gaming interface, which allows for a seamless and familiar experience across a field of users, while allowing each user to customize a product based on their own specific goals, budget, and living patterns.

Unlike an online car configurator, ZHA had a goal that users would interact with their environment in a way that information was built up, consistent with a three-dimensional modeling package in which more basic geometries, such as points and lines, are then expanded into surface geometry and given *thickness*. For ZHA's configurator, this meant that users started with the most basic of digital elements: *pixels*. In the spirit of gaming, the configurator allows for competition, so users can "bid" for the same pixels, which in turn drives up the cost. This allows for a digital equivalent to the kind of competition regularly seen in a speculative real estate market. In this gaming environment, the design team found that in initial tests, all users chose the pixels or "lots" at the corners, passing on the pixels that would constitute "mid-block" sites. The idea here is that users were selecting pixels that allowed for "viewing rights" without having to pay extra for them as corner views are less obstructed.

ZHA considered these moments of discovery as prescriptive in the way they would engage both the manufacturer, with supply chain constraints, as well as the end user; in each case the choices had financial implications. The initial "deployment" of the configurator is complete, with groundbreaking underway for the first 13 houses. Not only does this pilot project allow the firm to tailor its experiences to end users in a very precise, and needed, typology: housing; it also allows the firm to better align its design intentions with the free market, which is another goal. This shift to a more market-based product allows user choices that help drive creative logics. ZHA has refined its design ideas with respect to participatory co-authoring, but are keen to point out that these seemingly new directions have only been possible given the firm's ongoing involvement in commissions, such as museums and transportation centers, that have helped define it for the past 20 years.

Bhooshan suggests that there has been a "pipeline of projects" of a certain scale that allow for discovery. This occurs specifically in the deployment of complex geometry, which, when properly designed is "extraordinarily beneficial" in its capacity to solve structural problems or provide organizational logics that impact material volume or use. The tools the firm has developed for the more complex, unique buildings are now field tested, and have the capacity to be customized into a series of building components and applied at various scales within a *platform* design model. The firm sees this work as a natural step in its trajectory in terms of the ideological and creative positioning of its work. The notion of a platform, used broadly in manufacturing industries, encompasses commonality in design and production efforts, so that both assembly processes and physical components can be shared across a brand. An integrated platform model streamlines cost, fabrication, and assembly schemas though this repetition; while still allowing the manufacturer to customize certain aspects or components within the process, allowing for different models within the overall type or brand. Such a model grew out of the standardization practices of the twentieth century, while maintaining a scale of operations that allows more bespoke or mass-customized solutions. Platform thinking is central to new models of architectural practice and production, while giving a larger voice to individual users, traditionally our clients.

A conventional design and construction scope limits the interaction between the architect and those who have more downstream responsibilities in the building process. Such a siloed way of working has already been challenged by the board adoption of building information modeling technologies. ZHA, like SHoP, sees further opportunities through a platform-based way of working, which is consistent with the way 3D modeling and production tools such as CATIA are utilized within their modeling environment. For Bhooshan, at ZHA a platform design model has been adopted within a "prefab 2.0" strategy, where they can refer to a database of both physical suppliers and virtual design processes and components, allowing designers to work from and within an applied set of constraints while still having the ability to customize solutions. These operations are not intended to limit design teams but allow them to easily access the firm's body of previous work while speculating on new forms of novelty.

The notion of platform has also allowed ZHA to understand more directly its design efforts within manufacturing schemas while focusing more specifically on end users. This work has already been understood within a building's structure. Previously a structural solution for a column grid, rooted in ideas about standardization, would be sized for the largest load, then specified as a single size for individual columns within that field. Customization has allowed for difference in not only the specifying of individual components within the field, but manufacturing them in a cost-effective way, so that a grid of varied sizes could be produced economically. Material savings are a byproduct of such a process as well.

ZHA is expanding this type of thinking more broadly to other components within a building system, including floor slabs and walls, while still considering the end user. In a sense, they are moving toward an industrialized construction solution.

PLATFORM PROJECTS

A recent example of these efforts has been ZHA's Roatán Próspera modular housing solution for a small island off the coast of Honduras in the Caribbean Sea. Specifically, the platform and its larger organizational concept is applied to the more pragmatic and highly constrained ideas of prefabricated housing, with the intent of creating a level of customization across a site. Repetition, with some level of difference, is clearly rooted in the firm's efforts with geometric novelty. In producing the latter, the firm literally goes through hundreds of iterations, tasking the design team with selecting a solution(s) that best meets specific project criteria. Through the platform model, ZHA utilizes the generative capacity of difference, allowing it to be considered by others – specifically future fabricators and users of ZHA-designed buildings. Through automated and robotic manufacturing, as well as manual assembly, this group of users can select geometric schemas that work within the constraints of a specific situation. Constraint is applied productively through computation.

In this sense, there is a more targeted and sustainable aspect to ZHA's production and workflow – schemes that would have been previously discarded live to see use in what may be a wholly other situation. This *virtual adaptive reuse process* not only has implications for the previously discarded virtual information, but also for the assignment of both design and economic value to an aspect of the architect's work that had previously been less considered, furthering the architect's agency and future impact. Exploration, and exhaustion, of design concepts no longer occurs on a by-project basis but contributes more broadly to an overall vector or body of work. At ZHA, this body of work resides in what Bhooshan refers to as a *search space* which the firm can explore in an exhaustive manner and exploit in determined ways. Ultimately, solutions intersect in ways to satisfy multiple stakeholders. One could argue that such a process is one architects have always engaged; however, information holders have traditionally been more limited, and iterations more manual – a configuration is arrived at, engineered, revised, and so on.

Economic rationality figures into this workflow, in that the wisdom of the free market with a larger set of "rational actors" – suppliers and producers – allows for a more efficient allocation of resources. This thinking is not one of political economy, but more that the "market" has a greater capacity to satisfy the needs of end users. This also has sustainable, and creative, implications in buildings; their function may have a more transitional arc allowing for adaptive reuse considerations instead of a more conventional idea of demolishing and building again, something seen quite regularly especially at the interior scale of buildings. Here the *configurator* logic plays an important role. If various stakeholders can participate in a design and building solution there will be less impetus for change over time. This should not imply a design-by-committee type approach, however, as it positions the architect as the designer of a solution set, with a range of outcomes able to satisfy the organizational, structural, and programmatic needs of a set of users. Manuel DeLanda has explored these ideas in his thoughts on artificial intelligence within the architectural design space. Speculating about this phenomenon in 2006, he writes of AI and genetic algorithms, software that was "created not to aid designers but biologists in understanding the dynamics of evolutionary processes. . . . When used as a design tool, the designer's role is to decide, at each generation, which forms will survive and which will die, or in other words, the artist's role is to *guide the evolution* of these forms. In performing this guiding task, the designer becomes a kind of animal breeder: a dog or a horse breeder can hardly impose a predefined form on his animals and at best plays the role of aesthetic (or functional) judge."[2]

AESTHETICS AND THE DEVELOPER SPACE

The platform approach has also allowed ZHA to engage developer-clients, particularly in its ongoing efforts in China, where its volume of work has steadily increased over the last 10 years. The configurator has allowed the firm to proposition developers on different approaches to value and service in this context. The firm delivers value on the one hand through the increase in market price for leasable space in a ZHA building, which they justify through simple economic demand, suggesting a developer should be committed to providing what an end user wants and what they will pay for in a given market. What is economically rational should satisfy both the consumer and investor/producer, with the producer being the building design and delivery team. At the schematic design phase, design teams ensure that development criteria such as zoning metrics and FAR, as well as client budgets, are met while defining a narrative that will bring people, and interest, to the building. It is the latter that can increase the market value of a ZHA building, in a way assigning value to a building's aesthetic criteria. The 1.3 million followers of the firm's Instagram account, and the visually competitive nature of social media, is not lost on Bhooshan. Although this sort of *brand value* is directly related to Zaha Hadid and the aesthetics put forth by her firm, it also functions as a literal *driver* – it drives people to the buildings where they can interact physically. Bhooshan suggests that this level of user engagement, first visual and then spatial/material, relates directly to their commitment to user experience and is reinforced by the firm's efforts in customization and manufacturing; and it is also consistent with a developer's model of fulfilling demand. Through an augmentation of the traditional design process by ascribing value metrics and constraints, both in a dimensional and economic sense, to more ephemeral aspects of architectural design, ZHA has found a means to satisfy more conventional development goals while creating the possibility of novel architecture, whether bespoke or mass-produced with difference, within the developer space.

GAMING THE WORKFLOW IN THE METAVERSE

The game engine repositions the end user in the design process, which has implications in the development of a computational workflow. Through the game engine, the audience of people engaging the work expands. The audience could be members of an augmented design-build team, as in traditional BIM, or users that will ultimately occupy space – loosely connected by the *social* interconnectedness of virtual environments. Such connections are not ambiguous to the firm, as understood

Figure 6.2 Roatán Próspera Residences, Zaha Hadid Architects, 2021: Referring to their work with the game engine as a "platform for people," Zaha Hadid Architects is exploring an intuitive interface to allow immediate feedback from user-based input.

through their social media presence. Bhooshan sees the work expanding to a fully realized metaverse with its sociopolitical, economic, and cultural delineations as real as in the *actual* world.

The studio continues to experiment with the game engine, pairing the user participation it fosters with easy-to-use tools and sustainably produced building parts. In the Roatán Próspera pilot project, the firm is venturing into mass-customized and modular houses off the coast of Honduras with a series of island homes currently being developed by ZHA with the multidisciplinary engineering firm Hilson Moran. This system is unfolding in the broader metaverse, served by both increased participatory possibilities and heightened user-, social-, and spatial-experiences.

The game engine comes from a general idea of a "technology stack" to host communities "both online and on land." It is based on the idea that geographic boundaries have become increasingly blurred and all the technologies that are needed to host communities *spatially* online are now available. Bhooshan's team at ZHA has always been interested in computer graphics and geometry and has found modeling primarily used for building design explorations to have commonalities with technologies used in game design. Bhooshan points out that CAD software does not represent users, and while more advanced BIM technologies can give building elements relational characteristics, these are set by the designer and not open to more chance encounters like those within a video game, which are built around user choice. The firm's work in robotics and digital fabrication

has likewise expanded, so that full-scale parts for larger building assemblages are possible. Bhooshan sees this confluence as "fueling the metaverse."

Central to ZHA's approach to sustainability is the reduction of material and the amount of energy needed to produce, assemble, or maintain its buildings. In the case of Striatus, its precast modular prototype bridge, strength is achieved through geometry, as opposed to reinforcing or a high-strength concrete mix. Standard strength concrete typically uses less energy in its production and has less embodied carbon than enhanced mixes. In this sense, it is a funicular structure, relying on a desired resting state of the material for its form. While the bridge does have some steel parts, these are discreet components that connect to the concrete. This separation of material means tensile and compression functions are clearly delineated per component and makes the structure more fully recyclable.

ROATÁN PRÓSPERA RESIDENCES

The firm's modular housing proposal, Roatán Próspera, emerges from a long-term goal of changing how architects design and build using computer graphics tools, which are increasingly informed by digital research that is applying more traditional wisdoms located in ancient materials such as masonry and timber with automated and robotic processes. Before automated technologies, these construction methods embodied certain geometric agencies that gave them force-flow logics and resistance – the ability to hold form and transfer load. These were

Figure 6.3 Striatus, ZHACode, 2021: The inclusion of more analog material principals with a heavily automated design-build workflow was explored by ZHACode in Striatus, a concrete-printed prototype for a modular precast concrete bridge. The span is free of tensile reinforcement – there is no fiber or steel in the concrete. The sections are hollow and join through friction, making it easily assembled and taken apart.

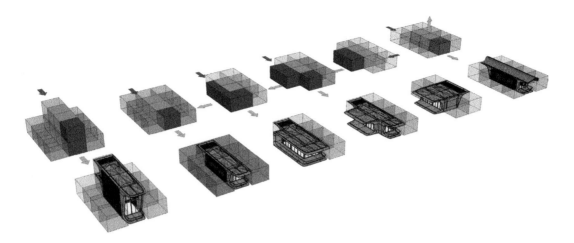

Figure 6.4 Roatán Próspera Residences, Zaha Hadid Architects, 2021: User input – choice – was originally configured with a pixel organization schema based on 12-unit grid 3 pixels wide and 4 pixels deep. This proportion related to a fairly dense land-use plan, with semi-dense houses aligned with frontages shorter than they were deep.

also energy-conserving paradigms that had a natural basis. The incorporation of these properties with more automated construction processes maintains the familiarity with such techniques but also addresses supply chain and labor issues.

Multiple computational techniques were used to adapt modular technologies to the steep and ungraded terrain and climate. This served the goal of shipping standardized units to the remote island. User choice was also considered early in design, starting the input process with a 12-unit grid with each square representing a pixel. This grid was initially simple block geometry that, once activated, could be modified three-dimensionally to create an envelope. Pixels were color coded so that the resultant tiling pattern could take on program, which initially served to subdivide pixels into indoor and roofed spaces, and outdoor and patio spaces.

Designing for choice, as is customary in gaming scenarios, is somewhat different in architecture where solutions are preferred to be novel and bespoke. At ZHA Bhooshan and his team had to anticipate a range of solutions that users could activate within the same set of tiles – the resulting typology would be a sort of articulated bar, which historically applies well to multifamily housing. Such a workflow proved challenging for engineering consultants, who typically expect plans to be "frozen" or relatively set so that "backgrounds" or a model could be transmitted for development. In other words, engineers typically solve singular solutions within a conventional design development process, they don't solve for a range of options. This captures a significant aspect of future design and building – that architects and our engineering partners should no longer simply *solve a problem*. Instead, we should be identifying issues and opportunities, seeing things for what they are, withholding an immediate or local judgment, and applying a longer-term type of design thinking that allows for hyper-articulation of a concept.

This brings up considerations for the contractual relationship between architects and engineering consultants, and the client – the end user who will ultimately live in the residence, as well as manufacturing and assembly.

The firm returned to the idea of the circular factory, in this case, calling it a *micro-factory*, which was set up for production in a way that raw materials and physical effort – the traditional mode of construction on the island – were mated with a more automated basis. There were specific temporal considerations given to the receiving of materials and the time needed to process them into building products. Through the firm's "configurator" interface, users are

Figures 6.5, 6.6 Roatán Próspera Residences, Zaha Hadid Architects, 2021: ZHACode's "configurator" is an intuitive online interface that allows users – people interested in purchasing and living in the residences – to assemble their houses based on a series of options. Larger organizational metrics, such as levels and height, and surface orientations – flat, sloped, are input as are spatial resolutions that give way to terraces and interior spaces. Through this virtual building process, users toggle between site-specific visualizations to understand orientation, more abstract views to organize program and space, as well as rendered interior visualizations of rooms. With the latter, interior views confirm internal configurations as well as views from the house to the surrounding beach and landscape.

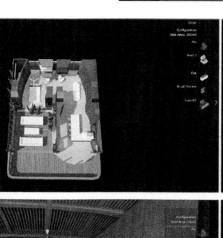
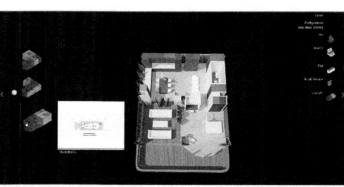

not only able to configure interior organization based on program, orientation, and view; but can stack the pixels, or building blocks, so that these orientations can be optimized per user preference. Raw materials and time were understood first through the user-generated geometric organizations, and then through a transformation process to actualize the building components by the architects.

More globally, individual user input is visualized adjacent to other user choices and common amenities, developing the community virtually as geometry is prepared for further translation. The concept of air rights, or the volume of space above the two-dimensional plane of the residence, was also considered. The level of satisfaction delivered to the home buyer is more instantaneous and satisfying, as a fairly "complete" visualization can be arrived at once the gaming interface is understood. This again is a telling trait of building futures – the idea that design intentions can be better visualized more quickly with more immediate user input mediated through a virtual environment. Recall, though, that the architect, as author in this sense, needs to anticipate and provide for options, building variation into the design process itself.

Configurator sessions occur with buyers where members of the design team are present to understand the rationale of user choices. Through what the firm refers to as a "pilot phase" they are working with buyers and learning about their choice making, which in turn allows for adjustments to be made in the game engine. The goal is for the game engine to be a completely discrete object that lives online in the expanded metaverse. The process is initiated through contact with a business called Próspera – many developers create discreet limited liability companies for singular development projects – with a sales team as well as members of ZHA and ZHACode. This entity then licenses the designs, which includes the configurator, which remain the intellectual property of Zaha Hadid Architects. Designs are localized within a "founder's village" – the site of the residences – through this entity, which also interacts with the buyers and organizes scheduling and configurator sessions.

For the operational environment of the configurator, ZHA imagined it would function like an online game, where user input is largely provided via mouse clicks. Such a goal precludes prior knowledge of the game environment and allows for a more democratic format so that "more experienced" users do not have an edge in selecting residential options. Users simply click on the pixels. Bhooshan allows that within the three-dimensional configurator environment navigation – not selection – has been the biggest challenge to the users. The firm is less concerned with this, since they have been present during

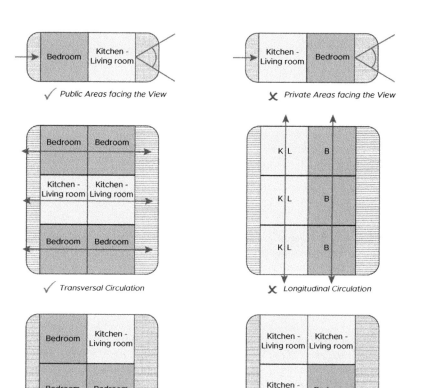

Figure 6.7 Roatán Próspera Residences, Zaha Hadid Architects, 2021: A series of intuitive manipulations of the original, and inert, pixels allow user goals and a geometric intelligence to quickly be added to the tile field. Choices such as public-facing views, private spaces like bedrooms, and transverse circulation between spaces can be chosen.

the selection sessions with the buyers for the first 13 dwelling units, with an understanding that modifications can occur based on the results of this process. Of more interest is how people are justifying choices within the game engine. Most buyers have immediately navigated to the book-end position of the units – the units on either end of the bar – and many navigated away from a lower-level residence, even though there was direct walk-out access to exterior space. These "popular" selection choices were noted by the team, which found there was a great degree of overlap after the first round of the selection process, with multiple users selecting the same pixels.

Choices were segregated into color-coded three-dimensional volumes so that ZHA designers could "recompose" these clusters to maintain user choice while allowing for a more singular global authorship imparted by the architect. They referred to this more manual modeling phase as the "reconciliation phase" of the configurator. Conflicting choices in some cases were resolved geometrically, but in others an overall market logic was applied where a more optimal configuration went to the highest bidder. Bhooshan sees these local trade-offs as positive, especially since individual choice and benefits were largely fulfilled in the process. The design team further subdivided the clusters which expanded the footprint of the development while

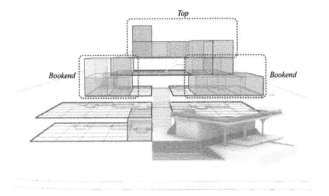
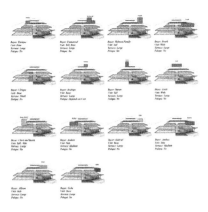

allowing more users to have peripheral and end views. Preferences to interior layouts were also noted, with there being similarities between the relationship of more public spaces like kitchens to the rest of the house, as well as the overall proportion, which remained related to the original tile pattern.

Following a local configuration process initiated by individual users and the more global coordination of the emerging clusters, the architects could go back to the project engineers having significantly limited the amount of variation to engineer. This reduced the number of conflicts that needed to be resolved by the structural and MEP design teams and allowed ZHA to locate fire-rated separations for life safety. The multi-family configuration made for common spaces and shared building components such as wall and floor assemblies to

Figure 6.8 Roatán Próspera Residences, Zaha Hadid Architects, 2021: Once user input was received, the architects had the opportunity to integrate user choices, performing refinements to the global configuration of the community while keeping the essence of all user input. This work resolved overlaps – volumetric clashes of geometry based on user choice. A bookend strategy was arrived at through this process so that many of the 13 units were able to be opened to the side-yards, allowing for "free access" to guaranteed views of the tropical surroundings.

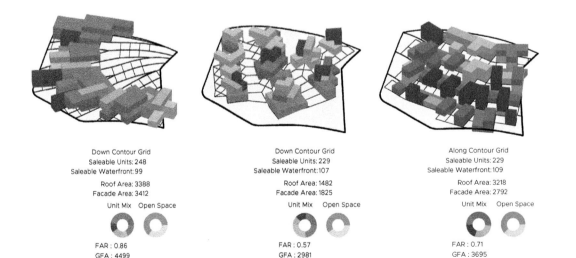

Down Contour Grid	Down Contour Grid	Along Contour Grid
Saleable Units: 248	Saleable Units: 229	Saleable Units: 229
Saleable Waterfront: 99	Saleable Waterfront: 107	Saleable Waterfront: 109
Roof Area: 3388	Roof Area: 1482	Roof Area: 3218
Facade Area: 3412	Facade Area: 1825	Facade Area: 2792
FAR: 0.86	FAR: 0.57	FAR: 0.71
GFA: 4499	GFA: 2981	GFA: 3695
Net Saleable: 83% Open Space: 35%	Net Saleable: 82% Open Space: 40%	Net Saleable: 96% Open Space: 38%

Figure 6.9 Roatán Próspera Residences, Zaha Hadid Architects 2021: Individual buyer choices were captured through the game engine and then recomposed by the design team to allow for the maintaining of individual choice while bringing a more specific site and clustering response by the architects. This workflow allows for a "both/and" condition where users receive their preferences but there is still a global logic to the development, which is imparted by the architect. These ideas about authorship have been discussed by Manuel DeLanda and others regarding architects designing with generative algorithms.[3]

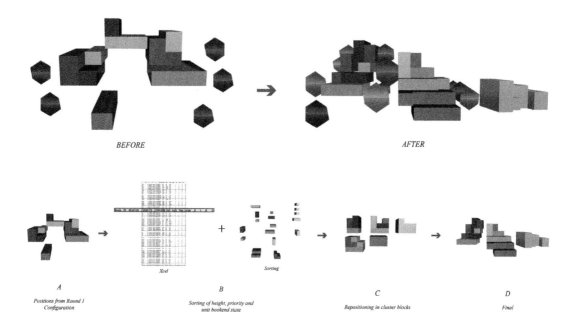

Figure 6.10 Roatán Próspera Residences, Zaha Hadid Architects, 2021: Following the first round of configuration, where resident-participants selected "pixel" organizations of their houses, the Zaha Hadid team sorted the user-input geometry based on height, priority, and site location. Ultimately, the clusters were repositioned in a way that maintained individual user choice but created an overall formal organization that was more constructable.

contend with, as well as issues such as individual metering of power usage – common issues in housing typologies. These were also approached in a rule-based fashion in the application of building code and its compliance.

In a sense, the architects were lending their expertise at points in the process where individual choice was maintained but resolved against constraints buyers were not able to address. This sounds like something we have always done – a sort of "win-win" for the buyer. The quality of choice, which is the primary driver of the development, still belongs to the buyer, and the resolution of the compliance to these choices remain resolved by the architect. The distinction is that architects are working from a desired qualitative outcome first, and then applying code logics to it, as opposed to the other way around. Code becomes the lowest-common denominator of the units, with individual choice being the most expressive – goals that have been articulated further in some of firm principal Patrik Schumacher's writing.[4] It is also important to note that high wind resistance, categorized in relationship to coastal regions and the probability of significant weather events such as category 5 hurricanes, and seismic action, is understood within the design process. To engage these factors while being cost-conscious, the design team again looked to geometry, understanding how the interrelations and operations of the individual units could be applied to a kind of structural resistance. Shared areas such as hallways and amenity spaces become common – they are both communally owned pixels in the game engine and physical spaces.

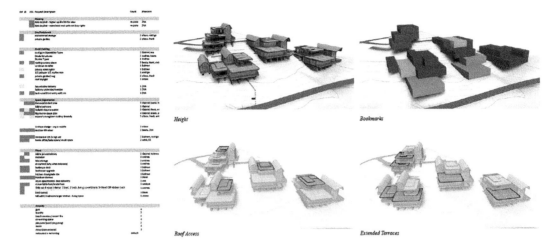

Revised floor plans were circulated to the buyers who, upon review, provided further comments which were addressed. This localized customization gives the project a variability while still allowing the architects to move the individual user choices into a more standardized fabrication and assembly schema. For Bhooshan, the resulting tectonics aligned with aesthetic choices across demographics and set up production methods such as 3D printing for masonry-type applications, or the use of timber in frame-based applications. Localized aesthetics could also change the look and feel of individual units while still relating back to the underlying game logic.

In moving these user-based geometries to production, the firm relied on its research into robotically made building parts, stemming from recent explorations like the Striatus Bridge. The goal is to grow these "technologies of consensus" to address larger and more contextually varied building problems, where consensus between all interested parties may not be a given. In this sense, housing – specifically, multifamily housing – becomes an increasingly valid exploration.

Once the configurator lives autonomously online, the potentials for scalability and diversity are immense, and relates to ZHA's broader ideas about the metaverse. Online users can "play the game" having a presence in multiple virtual communities, choosing to enter and exit with further choices for actualization or the making of physical properties. Scalability in this sense also means digital infrastructure, something the firm is exploring in terms of funding opportunities and partnerships. Interestingly, for all the required computing power to more broadly adopt such a design basis, the user interface remains simple and familiar. The configurator has been tested to work through most popular web browser applications such as Google Chrome and Microsoft Edge,

Figure 6.11 Roatán Próspera Residences, Zaha Hadid Architects, 2021: Individuals overwhelmingly choose to have their units "bookend" the unit assemblage – occurring at the edges of the building envelope. This allowed for one side of the residence to be separated from a neighbor. Within the configurator, a user could choose to buy "air rights" to maintain an open side to their unit, but following the sorting exercise undertaken by the architects, 88 percent of the users that chose the bookend configuration were able to receive their choice without an air-rights purchase – the architects reorganized the site based on choice while saving cost to individual owners.

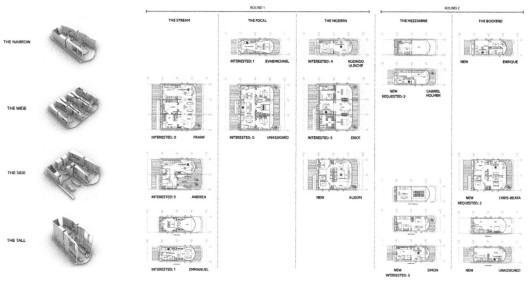

Figure 6.12 Roatán Próspera Residences, Zaha Hadid Architects, 2021: Typologies were standardized with naming conventions reminiscent of marketing trends. Models such as "the Modern" and "the Focal" were mated with spatial configurations – narrow, wide, tall – and then opened to a final round of user input. A running schedule of selection interest was maintained so individual buyers could understand their choices among others in the community while also understanding how these might affect the global configuration of the development.

and user selections are captured by a local data store. The backend of the configurator exists in the cloud, and uses Unreal Engine, a programable real-time 3D creation environment.

Geometric options were initially input by ZHA in Autodesk Maya with some engineering coordination in Rhinoceros. Maya as a digital tool is sympathetic to a game environment – its primary geometry is polygonal meshes and rendering occurs through texture mapping. It has native interoperability with Unreal Engine, where 3D geometries, assets, are compiled and published as a game. Within Unreal, "states" or various outcomes are hidden until activated – by clicking – by a user. Again, user choice in this sense is pre-computed, anticipating a range of selections. It should be noted all geometry is parametric – as is geometry in a conventional BIM environment – but not "live," meaning the game works by users turning on and off preconfigured states. In theory, geometric configurations could be computed in real-time as users input choices, but would require a more robust infrastructure. Herein lies the most significant difference between game- and BIM-environments, with BIM technologies, which are applied to one-off, bespoke sites and situations, geometry is live and has a history that allows for more circular and iterative interaction. In the game engine, where interaction and visualization are paramount to the user experience, geometric solutions are preconfigured with rendering occurring in real-time. This is how gamers can get immediate satisfaction by arriving at the end of a maze or blowing something up – the game designer has already thought of this outcome.

In Roatán Próspera, choice is highly curated. About 25 configurations are possible within the configurator, but given siting considerations and additions made in the "reconciliation phase" by the design team, the possibilities seem to be broader.

"DE-RISKING" ARCHITECT AND SUPPLIER SCOPE

Zaha Hadid Architects believes the configurator will ultimately lessen the risk undertaken by stakeholders within a construction endeavor. Each stakeholder within a design and build relationship has discreet responsibilities, for funding, design, etc. and each of these comes with some level of risk. The transparency of the configurator figures into a future de-risking of these relationships, with the clarity it affords encouraging further cooperation. Computer graphics and the actualization potentials of underlying geometry foster a "what you see is what you get" workflow. Developers or investors are assured because the configurator inherently facilitates a presale process, which in turn can facilitate "cheaper money" – cash being lent with less interest. Users are subject to dynamic pricing, where earlier purchases in the process can be discounted. Time is also reduced, with preconfigured organizations already being engineered, which in theory lessens the path to municipal approvals. This de-risking does involve an increase in architectural scope, not only in terms of design of multiple configurations but also in the authorship of the overall process, which is an agency-expanding endeavor.

Figure 6.13 Roatán Próspera Residences, Zaha Hadid Architects, 2021: The design team imagined the residences would be modularly fabricated and assembled with a local labor force. The platform model enabled the team to source raw materials, in this case, primarily timber-based sheet and structural members, and created a sub-assembly routine that could be accomplished offsite, with modular parts built and prewired remotely. Once site preparation is complete, the modules are stacked with final wiring and plumbing.

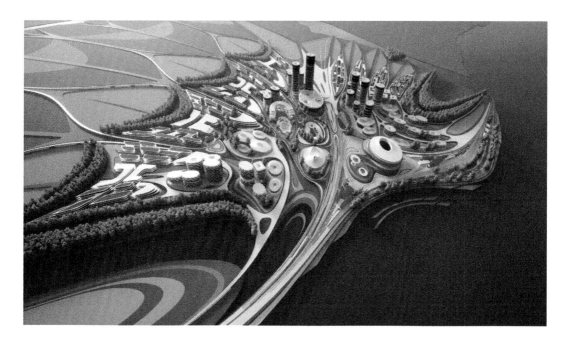

Figure 6.14 Liberland, Zaha Hadid Architects, 2015: Liberland is a micronation that has been operational online since 2013. The "country" has regions and neighborhoods, as well as an economy and functioning constitution, but possesses no physical footprint. It is both a conceptual and virtual representation of ZHA's goals for participating, and encouraging user participation, in a socio-economic virtual construction wholly existing within metaverse.

Bhooshan adds that both alienable and inalienable rights can be assigned to the pixels based on whether they give form to private or common aspects of the development, with some control over the amount of common space being created by the design team. This gives way to something like a digitally formed homeowners association.

This agency allows for a repositioning of the architect-developer relationship, which Bhooshan refers to as the "meta-layer" of de-risking, noting that many times developers make decisions based on financial (bottom-line) conditions, not understanding the total impact to design intent or quality. The game engine exposes a finer relationship between architect and developer, as well as buyer. Such a process could even decrease the ultimate scope of a developer, especially if their

Figure 6.15 NFTisms, Zaha Hadid Architects, 2022: NFTisms, a highly curated online exhibition by ZHA debuted in Miami in 2021. Based on the sale of virtual art, trade occurs entirely between virtual participants and is facilitated through the exchange of nonfungible tokens (NFTs).

main role is financial, as the configurator's initiation of common elements could lead to other funding schemas and cooperation between buyers, which is ultimately what occurs in conventional condominium associations.

ON TO THE METAVERSE

ZHA has a somewhat political ambition in the democratization of virtual space, with an online micronation it supports called Liberland serving to incubate cyber-communities and economies online prior to actualization on land. Broadly defined, the metaverse can be understood as having a range of outcomes, from a robust and wholly virtual environment to the digital twin, with its precise digital state of a physical object, and vice versa. Zaha Hadid's architecture in the physical world remains user focused, both spatially and sensorially, providing novel architectural experiences. Instead of the virtual always predating actualization, as understood in previous digital, and BIM, paradigms, ZHA sees an opportunity to expand real experiences with virtual ones, allowing each to inform the other. This seems like a logical progression, with there now being two-way communication between the virtual and actual, between immaterial and material states of objects and human interaction with them. The metaverse then becomes a place to further consider user interactions, providing novel experiences online within a spatial environment. ZHA embraces the openness of Web 3.0, or simply *Web3*, where ownership is decentralized with the goal of creating transparency and structuring relations between individuals as opposed to singular organizations.

NOTES

1 Conversations with Shajay Bhooshan in the support of this chapter occurred on September 16 and November 29, 2021.
2 Manuel DeLanda, "Philosophies of Design, the Case for Modeling Software," Verb Processing: Architecture Boogazine, Actar, ISBN: 9788495273550, 2001, page 140.
3 Ibid, p. 141.
4 Patrik Schumacher, "Libertarian Urban Planning – Prospective Regimes for Liberland," *Liberland Press*, February 2020, https://liberlandpress.com/2020/02/19/liberlands-prospective-urban-planning-regime/.

IMAGES
pp 94–95, 103–107, 109–115 © Zaha Hadid

7 | ON ECOLOGY II

NEW WAYS OF SEEING THE WORLD[1]

This used to be real estate, now it's only fields and trees . . .
— David Byrne,
(Nothing But) Flowers, 1988

In the period between the world wars of the twentieth century, Martin Heidegger was working on a series of essays that would ultimately be published in the 1950s and 1960s. Although not so apparent at the time, these essays, including "The Age of the World Picture," which was written in 1938, anticipated the eventuality that we would, someday soon, be looking back at the Earth, seeing the world from a vantage point outside of its boundaries – no longer as humans-being "in the world" but somewhere beyond it. Since *Apollo 8*, the pervasive rise of information technology has affected the way we work, interact, and collaborate, and *see* the world. The synthesis of digital design platforms with simulation and increasing access to data in the form of natural phenomena, ecology, and building performance has allowed contemporary architects to engage design and imagine construction practices more carefully.

This relationship to the world, and more broadly ecology, has the ability to radically reconfigure the architectural project, which has to do with new ways of seeing architecture, and our work, within the larger global environment – the world – that we inhabit. In the twenty-first century, this seeing, and subsequent thinking has transitioned from the "ever nonobjective to which we are subject" to "an object that stands before us and can be seen."[2]

Though building information modeling (BIM) and digital design and construction techniques have become mature technologies, their utility as virtual tools, and more specifically the utility of models – the material constructs of architects – have found further use beyond the conventional *efficiencies* supported in BIM workflows, especially within virtual and augmented reality schemas. This has to do specifically with ideas of simulation and part-to-whole relationships as they shift from traditional BIM uses to technologies that allow for new ways of seeing – that is, understanding the role and

Figure 7.1 Earthrise, William Anders, 1968: While there had been previous attempts to image the earth from space, most notably in 1966 by the non-human ATS1 satellite, the first, and most widely known, image of the earth captured by a human was by astronaut William Anders during *Apollo 8* in 1968. The image is the first to document Earth as an object, as seen by Anders and humans in general from beyond, captured through the camera shutter.

agency of architecture within the broader world, beyond specific sites and solutions associated with conventional architectural scopes.

In this sense, simulation moves from the scale of building modeling to ecological modeling, and part-to-whole relationships become more extensive and open, beyond the scale of building components, to take into account both broader – community and regional – scale as well as those smaller than the building itself. Digital twins and the virtual operation of models, both during design and construction and later, figure prominently in this expanded model.

Ecological thinking has become quite pervasive with stories every day in the popular press, news, and social media. These media generally convey a sense of what we might think of as "doom and gloom" in terms of the state of the world, especially as it relates to climate change, pointing to a more pessimistic future. Critic Donna Harraway takes issue with the term *anthropocene*, suggesting that climate change doesn't go far enough: "It's also extraordinary burdens of toxic chemistry, mining, depletion of lakes and rivers under and above ground, ecosystem simplification, vast genocides of people and other critters, etc." Haraway recalls images of the science fiction author Kim Stanley Robinson and his book *2312*, suggesting the period of time between 2005 and 2060 (which happens to be *our* time) would more aptly be called the "Dithering"[3] and may be too optimistic. But architecture is a *projective act*; we architects bring a sense of optimism to the future. By virtue of our work, architects are constantly thinking about the future, or specifically future conditions – possibilities we hope to

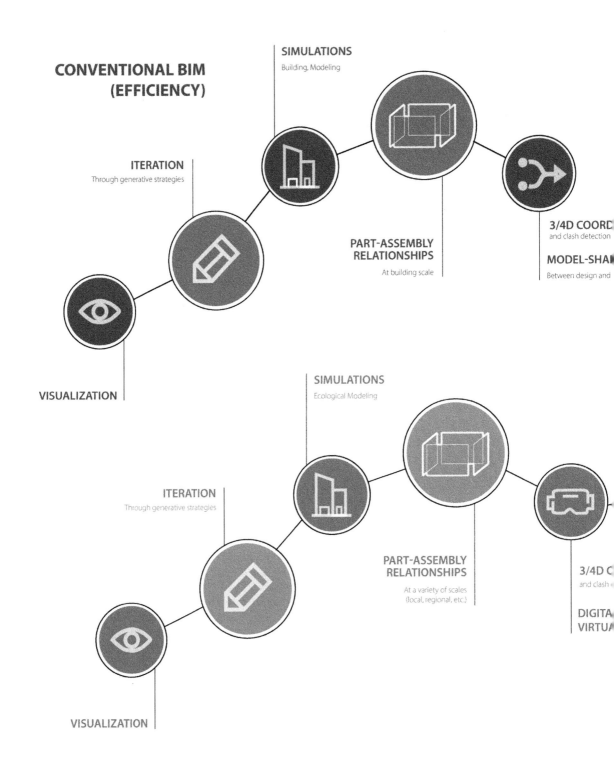

Figure 7.2 GRO Architects, From Efficiency to Seeing, 2022: conventional BIM schemas, preoccupied primarily with efficiency, are giving way, through new ways of *seeing* aided by technology, to a broader understanding of the work of architects in the world. In this sense, simulation at the building scale allows architects to engage with more broad and measurable conditions and with ways of understanding part-to-whole relationships. Allied scenario technologies, largely developed in video game environments, foster greater engagement with stakeholders in the design of buildings.

GAMING SCENARIOS

As a way of collecting input from various stakeholders, including community

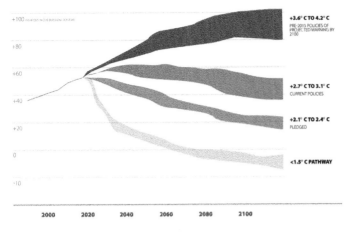

Figure 7.3 Climate Action Tracker, Pathways of global greenhouse gas emissions, published in the *New York Times* April 3, 2022: Recent advances to power generation and their economic accessibility have allowed humans to begin a kind of course correction, with current policies set to increase projected warming by +2.7° to 3.1° Celsius. The temperature has been reduced by almost a degree based on 2015 projections. Such a reversal has allowed for some cautious optimism in the human engagement of climate change.

actualize through our agency, which increasingly engages nonhuman conditions as we move to a more objective way of thinking about our work in the world.

Recent articles have acknowledged that humans have made progress in slowing climate change in recent years, noting that the cost of accessing solar and wind power has dropped considerably which has in turn decreased the cost of batteries – the latter allowing for an increase in electric vehicles being brought to market in the 2020s. According to German Lopez, writing in the *Times*, "The world was expected to warm by about four degrees Celsius by 2100. Today, the world is on track for three degrees Celsius. And if the world's leaders meet their current commitments, the planet would warm by around two degrees Celsius."[4] So humans have been somewhat impactful in the world, with current pledges reducing temperature rise slightly with a projected pathway to decrease by 1.5 °C, equivalent to 2.7 °F, from 2015 projections. Lopez's writing underscores the political commitment and cooperation necessary to make this all work, which is based on general agreement to target 1.5 °C at COP21 in 2015 with the understanding that a 2 °C increase above preindustrial temperature levels would significantly alter the world as we know it. Every tenth of a degree beyond the 1.5 ° goal makes climate damages that much greater and the overall crisis harder to contain.[5] Certainly, Latour's eco-criticism is timely here, and architects should be critical in how we assess and understand this data.

The American Institute of Architects (AIA) published in 2020 a Climate Action Plan, which they refer to as a "Climate Imperative." The report starts:

> Humanity is faced with a challenge unlike any we have previously encountered: we must take urgent action to reverse the *impacts of our* greenhouse

> *gas* emissions, protect our planet, and preserve life as we know it. Climate change affects every person, every project, and every client. The impacts are all-inclusive, with no respect for borders or boundaries—and are felt first and hardest by our most vulnerable populations. Rising sea levels, extreme weather events, and the degradation of natural resources are a direct result of increased carbon levels, threatening national security, global economies, and the health, safety, and welfare of local communities. Because more than 40% of U.S. greenhouse gases can be attributed to the building industry—during construction, embodied in concrete, metals, and polymers, and through everyday processes such as heating, cooling, and lighting—architects have the ability to lead the change our planet needs.[6]

The report is filled with urgency and is dire enough in tone that most architects will agree with AIA mitigation suggestions, including education of clients and municipalities, dissemination of best-practice case studies, and advocacy and activism in shaping environmental policy at the state and national level. These positions are relatively innocuous and align with general AIA policies of advocacy and education. Therein lies one of the barriers to operating in a more ecological basis: The steps we as architects – as humans – need to engage are somehow so obvious that our broad agreement and acceptance of them precludes immediate and local action, begging the question if we actually need a national advocacy group to guide how we should be acting in the face of environmental degradation.

The most pertinent aspects of the report are the four areas that fall within the scope of the architect's direct agency and ability to impact environments, especially in an understanding of energy consumption (and production) possible in a building or buildings; the ability through material applications to sequester carbon or generally lower energy use requirements; the ability to promote health, as an extension of health, safety, and welfare advocacy, in building design; and finally in the design of resilient and adaptable structures to anticipate future climate conditions and vulnerabilities.

Each of these, with perhaps the exclusion of "design and health," which has more immediate circumstances, falls within the projective scope of future possibilities and can be studied through the simulation capabilities of digital tools, especially when these are expanded to understand a level of engagement and impact beyond a building site. The suggestion is that through a more thoughtful mode of practice, knowledge sharing and visibility will occur.

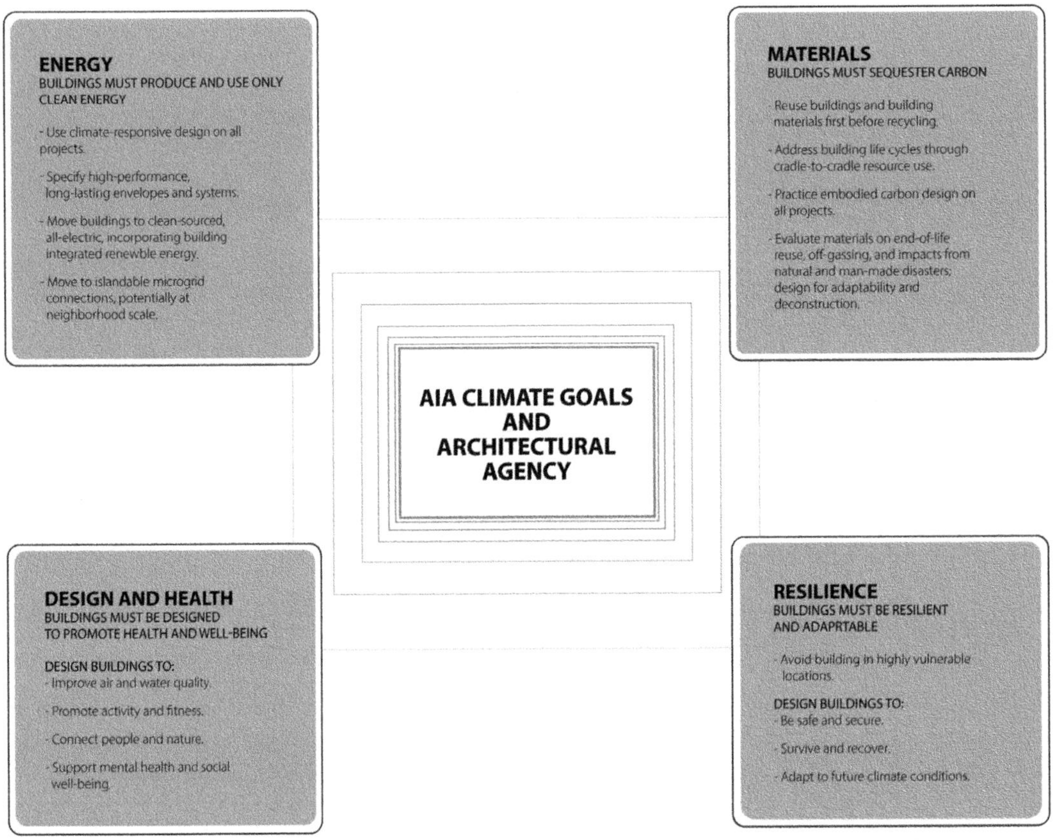

ENERGY
BUILDINGS MUST PRODUCE AND USE ONLY CLEAN ENERGY

- Use climate-responsive design on all projects.
- Specify high-performance, long-lasting envelopes and systems.
- Move buildings to clean-sourced, all-electric, incorporating building integrated renewable energy.
- Move to islandable microgrid connections, potentially at neighborhood scale.

MATERIALS
BUILDINGS MUST SEQUESTER CARBON

- Reuse buildings and building materials first before recycling.
- Address building life cycles through cradle-to-cradle resource use.
- Practice embodied carbon design on all projects.
- Evaluate materials on end-of-life reuse, off-gassing, and impacts from natural and man-made disasters; design for adaptability and deconstruction.

AIA CLIMATE GOALS AND ARCHITECTURAL AGENCY

DESIGN AND HEALTH
BUILDINGS MUST BE DESIGNED TO PROMOTE HEALTH AND WELL-BEING

DESIGN BUILDINGS TO:
- Improve air and water quality.
- Promote activity and fitness.
- Connect people and nature.
- Support mental health and social well-being.

RESILIENCE
BUILDINGS MUST BE RESILIENT AND ADAPRTABLE

- Avoid building in highly vulnerable locations.

DESIGN BUILDINGS TO:
- Be safe and secure.
- Survive and recover.
- Adapt to future climate conditions.

Figure 7.4 GRO Architects, after the American Institute of Architects Climate Action Plan, 2020: The AIA's Climate Action Plan identifies four areas in which it suggests architects can be positioned to engage and impact climate change. These areas, including energy and material considerations, are somewhat obvious, but should be considered with respect to architectural agency.

STATING THE OBVIOUS

All of this posturing begs the question, "What does climate change *look* like?" or, more appropriately, how can architects *see* and *respond* to something as variable and unpredictable as ecology relative to climate change? This question has conventionally been answered through a subjective, human, lens. As small changes to, for instance, weather patterns begin to affect human routines, those routines en masse have a way of affecting how we live in the world. In other words, much of the action that has taken place has been borne of *inconvenience*, or threats to the creature comforts of a capitalist (Western) way of life. Here it is useful to consider Heidegger's notion of *research*, which he suggests is "increasingly occupied by the figure of . . . the technologist," who only through the eyes of his age is "real."[7] Seeing what is *real* requires a shift to an objective way of understanding the world and our relationship to it, mediated, certainly in the practice of architecture, through technology.

The conceptual relationship of design to ecology has been considered by Timothy Morton, who in *The Ecological Thought*, writes:

> *The ecological crisis we face is so obvious that it becomes easy-for some, strangely or frighteningly easy-to join the dots and see that everything is interconnected. This is the ecological thought. Ecology seems earthy, pedestrian. It's something to do with global warming, recycling, and solar power; something to do with quotidian relationships between humans and nonhumans.*

In furthering this concept of interconnectedness, Morton relies on a redefinition of part-to-whole relationships that moves their traditional understanding away from *organism*, favoring a chaotic disaggregation of nature.[8] This idea of disaggregation, or separation into components, has recently been the mathematical basis for the study of dynamic systems, such as weather, studied via simulation.

Part-to-whole relationships also have parity in architectural thinking. The *whole* in architecture, a building or assemblage, is made up of a series of components that might be preconfigured or bespoke. Ecology, after all, is a Victorian creation, invented in 1866 by Edwin Haeckel to "formalize the principles of systemic interaction" that Darwin ascribed to nature in the *Origin of Species,* which reconceptualizes the idea of *encounter* within the natural realm. In doing so, Darwin's contemporaries responded creatively, according to Devin Griffiths and Deanna Kreisel in their work on Open Ecologies, "developing notions of ecology that were extremely wide-ranging and flexible, and applied to all interactions between living bodies, as well as their relation to the various organic and inorganic materials that constitute their environment." [9] For them, and like architecture, open ecologies are not marked by *despair*.[10]

Morton's work reiterates the fact that humans today are realizing that we are tied to, or in his words "surrounded, penetrated, and permeated by" nonhuman objects that we could not exist without. So this part-to-whole notion has resonance to our relationship to, and again not necessarily in, the world. Interconnectedness raises the level of systems thinking, and a systems-based approach to architectural problems, ones that beg we ask questions of these systems instead of rushing to provide solutions.

Indeed, the concept of organicism within the biosphere is based on the analogy of part-to-whole relationships within the human body. To some, Morton's holism is counterintuitive in that he contrasts the conventional belief that the whole is greater than the sum of its parts and that these parts are subjugated to it. Through this interconnectedness the whole has variability, to be understood as both less- and more-, a both/and proposition. In this case, the parts are not as easily replaceable; they matter in more significant ways. It is important to note that parts here are not simply components of a machine – a twentieth-century concept – but integral parts of an assemblage.

BEING NODAL

Nodal logic, half right, sort of wrong, and not entirely true, is not absolutist, but allows us to see the many shades of gray within a situation that is itself a contrast to the oppositional logic that guided much architectural thinking in the twentieth century. Ecosystems work this way; they are slightly incomplete beings or worlds in themselves. To demonstrate this, Morton conjures a blade of grass in a meadow. Removing a blade of grass does not destroy the meadow's "meadow-ness"; however, the continuous removal of blades of grass and the consequential flight of birds and insects, the incapacity of the soil through robust root systems to hold nourishment and water, and the inability to host various beings begs the question of whether it ceases to be a meadow – whether it ceases to be *real* in the way we *see* (understand) it. The meadow's meadowness – its being – is a form of radical coexistence grounded in interconnectedness of organic and inorganic things. This is what an architecture sympathetic to ecology looks like.

The figure of Morton's interconnectedness is the *mesh*, a multi-centered (centerless?) vastness with an anexact edge. The mesh is tiled, with higher levels of definition at its nodes, which give definition to ranges, more highly specific regions, within it. Here we can liken Morton's mesh with geometric ones. In *BIM Design*, a point-line-surface workflow was introduced and privileged as novel over predefined components. This was both a cautionary tale against the "completeness" of component libraries included in BIM systems as well as a call for novelty. Architecture as a bespoke project works this way: In setting up a more *specific* – not necessarily more *precise* – relationship in the development of geometric components, the designer enforces part-to-whole assemblies that have a level of robustness that is not reducible to either the thing or its individual components. It is something more than either, more than both, and has an agency that anticipates and thrives on interconnectedness.

In computer geometry, a mesh can define either a surface or a volume. Composed of a series of triangular or quadrilateral polygons, meshes, as extensions of

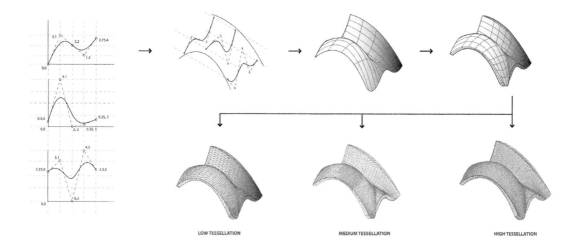

Figure 7.5 GRO Architects, Nodal Development, 2022: Following the point-line-surface assembly method advanced in *BIM Design* (2014), surface geometry is thickened to become a geometric mesh. These meshes are developed to convey architectural intent through broadened simulation schemes that take into account internal part-to-whole relationships, and scales larger than a single building such as the cluster.

predetermined geometries such as cubes, spheres, and cones, were downplayed in earlier texts about computer graphics in architectural design in favor of NURBS geometry, which writers such as Manuel DeLanda suggested were *more alive*[11]. Polygon meshes were initially favored by animators and renderers using polygon-based modeling programs, such as Softimage, 3DStudio MAX, and Cinema4D.

The notion of range, originally touted via the generative capacities of building information modeling schemas to yield a series of repetitive yet different solutions, is of further importance as we seek to define a more productive relationship to ecology and specifically an open ecology. In this sense, twentieth century utopian thinking gives way to a new type of optimism. Influenced by Marxist philosopher Ernst Bloch, Griffiths and Kreisel advocate for an ecologically utopian impulse that posits a "future-oriented hope for a better world"[12] found in cultural formations, such as architecture and other artistic forms of production. This certainly seems consistent with Morton's project, which seeks to link ecological sensibility to aesthetic practices.

SEEING ARCHITECTURE

As simulation shifts us to a kind of ecological modeling that leads to a broader understanding of part-to-whole relationships at a variety of scales, perhaps we can more readily engage the interconnectedness of things. A cluster can be defined as a group of similar objects or buildings and especially houses built close together on a sizable tract in order to preserve open spaces larger than the individual yard for common recreation.[13] If we think of our buildings not only as clusters of architectural components, which set up smaller scale part-to-whole relationships, but also as components within a larger field or environment, they become more like nodes, or the interconnected blades of grass in Morton's meadow, contributing to the

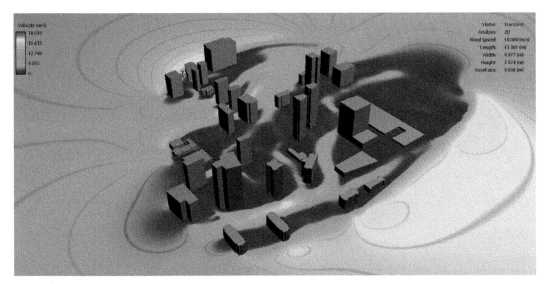

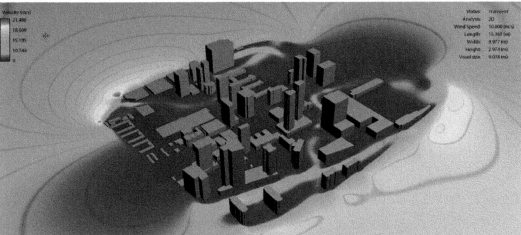

overall fitness of a larger system. In this sense, they can change based on various inputs. Being nodal and thinking about our buildings both singularly and as part of a larger meshwork is a shift from a more conventional way of understanding part-to-whole relationships.

In their Booking.com campus plan, discussed elsewhere in this text, UNStudio understands such a collective massing strategy in the consolidation of sites given the floorplate requirements of a larger programmatic commercial space goal. The architects utilize a mechanical approach called Underground Thermal Energy Storage (UTES) that operates at the scale of the urban block – a larger part-to-whole assemblage. Aquifers already existing underground are utilized for their thermal mass. Heat pumps pull excess heat from buildings, thereby cooling them, in the warmer months and pump this heat energy into the subsurface aquifers. The heat energy is drawn back out of the aquifers and pumped back into the building in the cooler months, utilizing cost-effective heat energy to warm

Figure 7.6 GRO Architects, Powerhouse Arts District (PAD) Wind Study, Jersey City, NJ, 2019: The simulation of environmental performance within modeling programs have led architects to rethink variables such as daylighting, air flow, and heat gain to understand more immaterial aspects of space-making, such as user comfort. Such tests were once given to specialized consultants and took long periods of time to generate. Architects can generate these simulations as part of their own design process with relative accuracy.

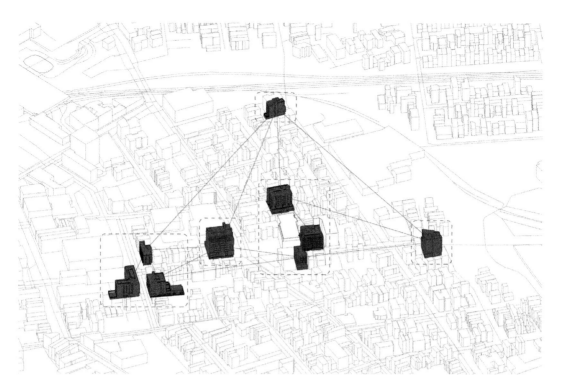

Figure 7.7 GRO Architects, Journal Square Cluster 01, Jersey City, NJ, 2022: Part-to-whole relationships of architectural components occur simultaneously at multiple scales, including ones through which we can assess the larger fitness of communities or environments. In this sense, buildings are understood as nodes within a larger mesh.

buildings. The solution is straightforward, cost-effective, and simple, and is a way that architects can operate at larger scales, understanding the ecological impact of building collectives.

In further outlining their call for open ecologies, Griffiths and Kreisel propose a "manyfesto," which they calculate to "open up . . . practices of naming, categorizing, and metaphor making . . . a sounding out of new approaches rather than an exhaustive catalog:

1. Open ecologies are situational: rather than focusing on a single actor, species, or stratum of the environment, they are defined by the interaction of diverse inorganic as well as living components.
2. They are compositional: they are not organic units or holistic cosmologies but instead involve multiple actors with differing interests.
3. They are nonprogrammatic: their forms are emergent rather than predefined or autotelic; their patterns and futures are unpredictable, chancy.
4. They are *abnatural*: they are characterized by uncanny interpenetrations of the manufactured and the other-than-human.
5. They are marked by uneven distributions of power; they demand that we reconceptualize modes of violence, from the environmentalism of the poor and the ecologies of race to the reframing of toxicity, threat, and predation.
6. They are neither preconcerted harmonies nor utopias.[14]

This ultimately leads them to autopoietic systems; that is, systems that are inherently sustainable by creating and maintaining their own parts. The term was first proposed by Chilean biologists Humberto Maturana and Francisco Varela to describe the chemistry involved in the self-maintenance of cells. Griffiths and Kreisel emphasize the system's cybernetic qualities of self-maintenance, suggesting that the part-to-whole relationship of an autopoietic system is its nature; contrasting this with a factory, which uses components to organize a system, a manufactured product, that is different from itself.

Their point is that open ecologies challenge notions of mainstream sustainability that don't square with a resource-intensive capitalist model of growth that ultimately depletes resources, furthering humans' complacency in anthropogenic climate change. Open ecologies are consistent with the expanded role of simulation in that they operate from a "vantage from which to see and calculate all the possible external costs (in energy, in carbon) that accrue with each decision about production and consumption."[15]

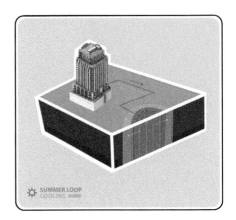

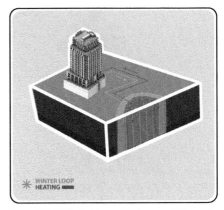

Figure 7.8 GRO Architects, Underground Thermal Energy Storage (UTES) Diagram, 2022: Larger-scale part-to whole assemblies, beyond the scale of individual buildings, allow for the pooling of resources at such scales to create effective heating and cooling schemas, such as Underground Thermal Energy Storage (UTES), which utilizes the thermal mass of underground aquifers to store heat energy in warmer months and draw it back up into buildings in cooler months.

NOTES

1 I would like to thank Cory Austin Knudson, who recently completed his PhD at the University of Pennsylvania and is the author of "Seeing the World: Visions of Being in the Anthropocene," *Environment, Space, Place* 12 (1) (Spring 2020): 52–82; and the students in my Seeing Architecture graduate seminar held at Penn in the spring of 2022 for helping to form these ideas.
2 Martin Heidegger, "The Origin of the Work of Art," *Basic Writings* (New York: HarperCollins Publishers, 1993), p. 143–203.
3 Donna Haraway, "Anthropocene, Capitalocene, Plantationocene, Chthulucene: Making Kin," *Environmental Humanities* 6 (2015): 159–165.
4 German Lopez, "Climate Optimism – We Have Reason for Hope on Climate Change," *The New York Times*, April 3, 2022.
5 According to Michael Oppenheimer, the Albert G. Milbank professor of Geosciences and International Affairs at Princeton University, quoted in https://www.kqed.org/science/1977423/straight-talk-on-temperature-heres-what-1-5-degrees-c-means accessed 5 April 2022.
6 AIA Climate Action Plan, published in July 2020 and available at https://content.aia.org/sites/default/files/2020-07/AIA_Climate_Action_Plan.pdf .

7 Martin Heidegger, "The Age of the World Picture," in *Off the Beaten Track*, ed. and trans. Julian Young and Kenneth Haynes (New York: Cambridge University Press), p. 64.
8 Griffiths, Devin and Deanna Kreisel, "Introduction: Open Ecologies," in *Victorian Literature and Culture* 48 (1) (© Cambridge University Press, 2020), p. 8, http://devingriffiths.com/wp-content/uploads/2021/08/0.GriffithsKreisel.Introduction_Open_Ecologies.pdf.
9 Ibid., p. 3.
10 Ibid., p. 17.
11 Manuel DeLanda, "Philosophies of Design, the Case for Modeling Software," in *Verb Processing: Architecture Boogazine* (Actar, 2001), ISBN: 9788495273550.
12 Griffiths and Kreisel, p. 18.
13 "Cluster," Merriam-Webster, https://www.merriam-webster.com/dictionary/cluster, accessed October 11, 2022.
14 Griffiths and Kreisel, p. 15.
15 Griffiths and Kreisel, p. 16.

IMAGES

pp 119 © NASA. Credit: William Anders; pp 120–122, 124, 127–130 © GRO Architects

8 | WINKA DUBBELDAM'S SYNTHETIC NATURES

Figure 8.1 Archi-Tectonics, Asian Games, Hangzhou, China, 2022: The scheme for the 2022 Asian Games, an invited competition, by Archi-Tectonics, with Thornton Tomasetti and landscape firm !melk was selected in 2018. The design team worked over the next four years to implement their vision for seven buildings within a 47-hectare Eco Park during fluctuations in labor markets, significant increases in the cost of raw materials, and a global pandemic. Through efficient use of BIM technologies, the design and contracting team were able to save 1,130 tons of steel and shortened construction time by 20%.

Winka Dubbeldam, a Dutch architect, originally trained in sculpture and architecture in Rotterdam prior to attending the post-graduate architecture program Columbia. She has been committed to working with manufacturers to develop more innovative building and façade systems to ensure that her design goals can be implemented. She formed her practice Archi-Tectonics in 1994, and the firm has steadily employed 12 to 15 designers since.

Archi-Tectonics' rich history of manufacturing engagement goes back to early commissions, such as its 80,000 square foot building at Greenwich Street, in 2000, where folded glass panels, manufactured in Spain, were assembled as a unitized system in just three weeks on site. The project also served as an early example of a digital workflow, where three-dimensional computer files were transmitted to the glazing manufacturer, a subcontractor that produced custom-extruded aluminum mullions. The façade's varied directionality lets residents more readily experience weather conditions such as a soft rain drumming on the diagonal glazing panels, and provides more views to the sky than a conventional façade. The late-critic Herbert Muchamp referred to the building in the *New York Times* as a "para-building" in which Dubbeldam "crystallizes urban complexity within the discrete architectural object."[1]

Archi-Tectonics has stated that innovation in architectural design and its structures requires a revolutionary change in the thinking of how architecture is conceived. They have reimagined the notion that a building is composed of standardized elements such as columns, floors, and walls, in favor of assemblies of mass-customized generative components. These components are more "organic" and resemble the human body in its complexity and natural fit. From a construction standpoint, smart components require a level of prefabrication, but prefabrication as pure repetition of standard elements is an outdated mode of operation; mass-customized components are evolving as a series of heterogeneous elements, defined by an analysis of specific performance requirements. A component's design intelligence is more akin to automobile and aircraft design than to conventional architectural design, and is more systems-based. This more systematic way of thinking is common for scientists and industrial designers but is relatively new for architects. Such operations are not only changing the way buildings are designed, manufactured, and assembled, but essentially allows an innovative path for architecture to develop in years to come.

Another way to frame the work of Dubbeldam's office is in the scale of buildings it undertakes. While remaining a generalist practice, the firm understands that innovation can be evaluated and prototyped more easily in smaller projects, including exhibitions and objects. The work varies from these "proto-projects" to the larger scales of buildings or master plans. Dubbeldam refers to the flow of ideas between their projects as "open source."[2]

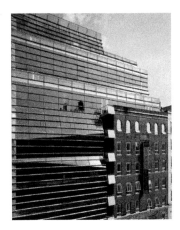

Figure 8.2 Archi-Tectonics, Greenwich Street Building, New York, NY, 2000. Folded double glazing with a solar film allows the glass façade to perform specifically for the New York climate. The glazing allows for solar penetration from the low winter sun, providing passive solar energy, while the solar film protects from the harsher summer sun, removing solar glare. Such an example of form-as-performance, articulated by Manuel DeLanda and others, was an early goal of architects engaging the digital.

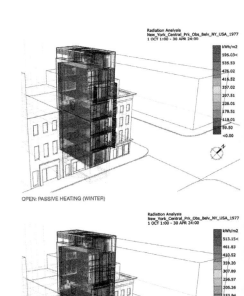

Dubbeldam considers the consequences of the anthropocene the largest and most important project we will work on as architects. Human activities have impacted the environment enough to constitute this distinct geological change, in which human subjects have become the dominant influence on climate and the environment. The United States, which contains about 5% of the total human population on Earth, alone uses 25% of the world's resources, including 25% of oil and coal and 27% of natural gas, causing 30% of the world's pollution. Still, recycling and composting have prevented 85 million tons of material from being deposited into the waste stream, up from 18 million tons in 1980 – a 470% increase.

Though we are slowly and collectively realizing that steps must be taken to address these metrics, Dubbeldam, like many architects, feels that we can do more, especially since the consequences, such as flooding due to increased rainfall, are being experienced at more regular intervals. These climate events have displaced thousands of people and caused billions of dollars in damage to private structures and public infrastructure. The costs associated with Hurricane Harvey, in the fall of 2017, are approximately $125 billion, second only to Hurricane Katrina in 2005.

Dubbeldam also believes municipal governments need to better acknowledge climate change by supporting an increase of flood zones and climate-event preparedness in general, which in turn would mobilize more architects. She believes "in this next wave of the anthropocene where

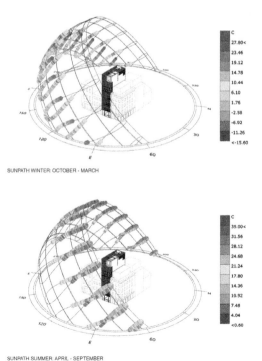

Figure 8.3 Archi-Tectonics, 512 GW Townhouse, New York, NY, 2019: Environmental plug-ins for Rhinoceros 3D have enabled architects to better understand weather, or the effects of local climate on their buildings. Dubbeldam and Archi-Tectonics use such tools to further simulate façade performance. A façade system of operable panels, a 3D trellis, creates a micro-climate around their 512 GW Townhouse project, allowing for shading, natural ventilation and exterior spaces, all wrapped in the 3D trellis volume.

climate change and environmental challenges threaten our and other species' existence, architecture needs to transform into a pro-active productive system, a body that has agency on its environment." Buildings will need to learn from nature's intelligence, by hybridization and symbiotic relationships with natural systems, becoming "synthetic natures" themselves. Buildings with such agency can foster positive feedback by, for example, absorbing carbon and emitting oxygen to their environment, and cleaning and filtering air and water. In this sense, Dubbeldam feels the role and scope of the architect has expanded, and the impact of our profession can become even more important to both ecology and building. She sees this as a challenge to architects working in the twenty-first century, which positions the projective optimism of the architectural project against the pessimism associated with much eco-criticism today – an opportunity to both "help and cause change." To this end, since the early 2000's she has thought pro-actively and holistically about buildings and building environments, adopting strategies for the adaptive reuse of structures or materials, and seeking siting solutions that can minimize energy use and maximize energy savings, either through passive solar heating or insulation. These ideas have also fostered early, and continuing, explorations with simulation software for climate-related study.

2022 ASIAN GAMES PARK, HANGZHOU, CHINA

Despite its boutique size, Dubbeldam's firm is increasingly becoming involved with larger-scale work. In 2018, the firm won the Asian Games 2022 competition for a 47-hectare Eco Park, a 1.6 km long green space containing seven new buildings, including two hybrid stadiums.

By understanding that the digital realm blurs boundaries between disciplines Archi-Tectonics proposed for the competition an equal responsibility-sharing team of experts, as opposed to a conventional organization of the architect as lead with engineering and landscape consultants. In collaboration with structural engineers Thornton Tomasetti and Dutch landscape design firm !melk, they not only felt this team of equals would be better positioned to win the competition but also believed it is an important model of future collaboration, where group intelligence is more relevant than that supplied by each firm, applying a part-to-whole approach.

The park design by !melk also implements what Dubbeldam refers to as a "sponge city" strategy by incorporating and restoring wetlands, creating islands in a waterway that bisects the site, and through the introduction of porous pavement that will enhance the site's hydrology. A reintroduction of local vegetation assists in restoring the natural biome. By focusing on the landscape structures built in the site, Archi-Tectonics' scheme not only conserves nature, but creates new hybrids that form symbiotic alignments between organisms and inorganic matter, assigning the latter organic qualities and a more significant ecological agency.

The two primary structures within the park, a field-hockey stadium/open-air cinema and table tennis stadium/concert hall, are both designed as smart, hybrid structures. The former includes a 125 meter free-span roof; and the latter, the intersection of two ellipses, features a massive suspendome roof, which makes the building adaptive and column-free. This allows the buildings and green spaces to transition from the Asian Games into a park and multi-use structures for the city of Hangzhou.

Dubbeldam recalls the design process, with decision-makers in each of the companies coming together to examine what she feels is one of the most interesting problems we face: creating a "green lung" in the middle of a city of 10.36 million people, based on a 2019 estimate. The hybrid structures for the Asian Games are developed with the city's legacy in mind. By understanding the stadia's fundamental duality, which allows them to function as a concert hall and event spaces for Hangzhou after the games, the buildings transform the park into a constantly evolving, dynamic environment, rather than creating static, representative objects that would become "white elephants" after the two-week Asian Games contests. Learning lessons from Beijing in 2008, with its structures largely unused today, Dubbeldam felt strongly that the park should have an anticipated purpose following the games. Looking at a wide array of potential users, the park would be both ecologically sustainable and programmed to ensure its continued use in the future.

Figure 8.4 Archi-Tectonics, Asian Games, Hangzhou, China, 2022: The site strategy provided by the client was rejected by the design team, favoring one that allowed both a formal and programmatic linkage between two stadiums planned for the site. A green pedestrian valley, with retail spaces on each side traverses the site at the lower level, connecting the stadiums via an outdoor walk and green space with green-roof glassed façade pavilions and long solar-paneled wings that shade visitors.

The site was bisected by a road and waterway at its center, and the competition required the two stadium buildings to flank either side. This conventional planning approach would have organized two halves of the park as the rear yards of the stadia. The team studied a series of other options and landed on one that pushed the stadium structures away from the center of the site and created a green "shopping valley" – a pedestrian mall that linearly connects both stadia while going *under* the street and waterway that bisect the site. While the design team felt this was the best site organization, they recognized that the massing strategy contrasted the stated rules and the planning goals initially set forth in the competition. The risk paid off, and the design team's concept, which better activated the park and utilized the stadia as attractors that set up the retail corridor, was endorsed – the competition sponsor and various government officials understood both the strategy and the passive environmental opportunities it afforded.

EARTH-BUILDING

Dubbeldam recognizes that crossing under the body of water and an eight-lane road required intensive civil engineering, but the significance of the site, and event, was important to the client. The result was a 503-meter-long shopping valley, allowing the retail units themselves to be nested into the earthwork, providing both insulation and green roofs. The bisected park is also connected by two pedestrian bridges. On one end the arcade terminates at the field hockey stadium, which is also used as an outdoor concert venue and cinema. The building is technically notable for the 125 meter free-span roof, which overhangs and provides shading to the 5,000-seat arena. At the other end is sited a 35,000 square meter hybrid concert hall and table tennis stadium.

As the overall section developed, so did the nesting strategies for the buildings set within the site, these became hybrid landform/buildings. The zero-earth strategy allowed excavated earth to be relocated on top of several of the buildings that were sunk into the ground. Skylights, which provide daylighting as well as ventilation, are playfully dispersed within the parking and other underground structures, creating a visual relationship between the earth and the synthetic constructions below it. The pedestrian bridges contributed by !melk, the landscape architect, allowed for overall vantages of the site as well. These bridges received LED lighting, which are powered by the photovoltaic arrays on "solar wings." Other important structures, such as the boomerang shaped exhibition center, were also nested in the landscape, allowing for relationships between buildings as well. This structure is essentially underground with two facades that emerge from the landscape and sit in front of the field hockey stadium.

The recently completed design had a strict government stipulation for 85% park, quite the challenge to place

Figure 8.5 Archi-Tectonics, Asian Games, Hangzhou, China, 2022: The massing plan nested buildings within the rolling landscape, with wetlands utilizing vegetation that naturally filters nearby river water. Small archipelagos were introduced to speed local water flow, assisting with the natural filtration, and increasing oxygen content. Indeed, non-human species were included on the design team's list of "new users." Once the site strategy was established, relationships between earth and building features were explored.

Figure 8.6 Archi-Tectonics, Asian Games, Hangzhou, China, 2022: A diverse group of human and non-human users were considered for the site, a concept that guided both ecological strategies as well as post-games use. Incumbent in this was an understanding of legacy, both for the Asian Games and Chinese land development, and a dynamic user environment that included green spaces, as well as active and passive event spaces.

the required seven buildings in the landscape. The team's "zero-earth-strategy" reuses excavated earth from reconstituted wetlands and the new shopping concept, called the Valley Village, to create a hilly landscape that effectively dampens urban sound and creates a serene park landscape. As buildings merge with the landscape, their 64,000 square meter combined green roofs act as an agent of environmental change by releasing an estimated 83,408 kg of oxygen and absorbing 114,846 kg of carbon dioxide per year, according to Dubbeldam. Understanding that the entire assemblage had to perform comprehensively as a "synthetic nature," the design team was very interested in reactivating local passive land strategies such as wetlands and bioswales. All walkable surfaces are paved with porous pavers, ensuring the "urban sponge" functionality. There is also an overall drainage and water-carrying schema, and many of the surface treatments are designed to allow for the absorption and recharge of stormwater – water management was key to the site organization. The team also created a series of archipelagos within the waterway, which helped control speed and flow, while providing natural filtering of the water. This drains off excess stormwater, while maintaining water on site during dry periods.

The buildings partially function through their connection to the surrounding nature; water is extracted from newly reconstituted wetlands, preserving energy while cooling the seats of the stadia. Natural ventilation and lighting are maintained through operable vent windows and a central skylight – the buildings literally breathe.

Understanding the post-games use of the site influenced the ecological strategy as well, with a pool of both human and non-human users being considered. Outdoor recreation features, such as a skatepark, playgrounds, and outdoor seating, are seamlessly integrated within the landscape. The rolling configuration allowed for natural shading, air movement, and sound attenuation, so spaces for reading or meditation are as readily accessible as playing fields.

EARTH-BUILDING, BY REMOTE CONTROL

Site and building construction occurred at the height of the COVID pandemic, with physical site visits being limited halfway through the construction period. The team relied on real-time drone flythroughs and weekly video conferences. Through the initial site visits, it became clear to the design team that some of the building intelligence developed during the construction documentation phase was better served by building information modeling (BIM), especially for the construction of the 125 meter roof of the field hockey stadium, and the table tennis stadium, the suspendome structure. The design team initially conveyed their intentions through three-dimensional models to a BIM team, after which design-detail sessions followed on Zoom. This significantly reduced costs, as well as material use and the project's overall schedule. Through this virtual process, 1,130 tons of steel were removed and construction time was cut by 20%.

Prior to the pandemic, the design team made several trips to Hangzhou to make the case for the technical solutions they proposed. On an early trip, Dubbeldam passed the site to find construction fences adorned with renderings of the project – construction had already started, enforcing what became a self-imposed deadline for completion of construction documents. This effectuated what became a kind of enhanced information exchange, fostered through modeling in Rhinoceros and Autodesk's Revit, and the Dynamo interface, on more complex aspects of the building form and structure, and even some landform aspects of the project. This workflow is notable for its optimizations, not only in terms of building components but in the means of construction.

Dubbeldam notes the idea of sharing is intimately linked to sustainability. Sharing implies a co-relation of humans and things as elements of one interconnected system. She believes a sharing economy is becoming relevant again today. As we continue to deplete natural resources, the need for reduction of production and waste – and for hybrid buildings that generate energy, clean air, and water – is increasing.

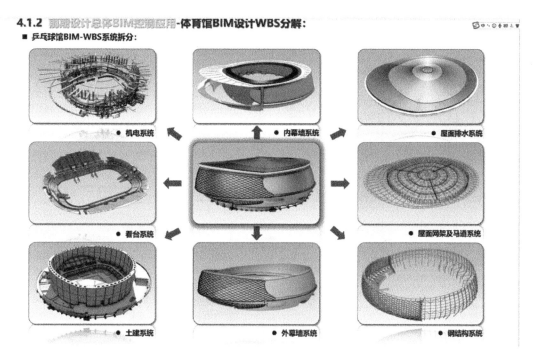

Figure 8.7 Archi-Tectonics, Asian Games, Hangzhou, China, 2022: Dubbeldam's team, in collaboration with BIM consultants in China, created a highly detailed drawing set that was transmitted directly to the general contractor. That drawing set, which details a series of cross-linear intersections that set forth the bounding dimensions of both stadia were critical to the successful delivery of the park and structures.

Figure 8.8 Archi-Tectonics, Asian Games, Hangzhou, China, 2022: In promoting ability over size, the contracting team deployed a fleet of smaller cranes which were quickly moved around the 47-hectare site. This ultimately proved more efficient than utilizing a single, larger, crane sized to lift the largest possible load. Dubbeldam feels this type of thinking greatly increased the speed of construction, ultimately allowing the park and its buildings to be finished ahead of schedule.

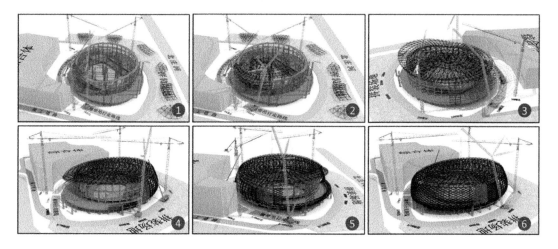

Figure 8.9 Archi-Tectonics, Asian Games, Hangzhou, China, 2022: Perspective showing the varied topography of the Asian Games Park supporting storm water absorption. Much of the planted earth was first bermed as part of a zero-earth strategy to redeposit soil mass within the site boundaries. The result was a gradually sloped synthetic nature – a hybrid landform/building that created diverse view corridors while passively channeling light and air within the constructed section of the site.

Figure 8.10 Archi-Tectonics, Asian Games, Hangzhou, China, 2022: The slope of the park, captured in a site section, allowed for a varied topography with structures nested into the landscape, thereby diminishing the perceptions of individual building heights and presenting an assemblage of both ecology and building. This synthetic nature is tuned to passively ventilate and drain, while holding the second largest multisport event in the world. Dubbeldam calls this synthetic valley the "inner sanctum" where there is complete quietness between the earth berms.

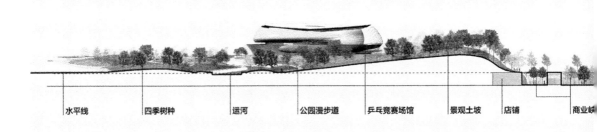

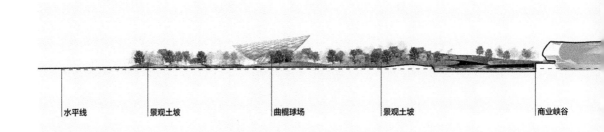

The extensive level of information sharing, including the electronic transmission of both two- and three-dimensional information, facilitated this collaborative approach and eroded the conventional division of labor, ultimately ensuring a clear understanding of intentions in the project delivery. Unlike a traditional construction project, for which the smallest crane possible to lift the largest loads would be scheduled for the shortest duration, the contracting team, aided by temporal simulation, deployed multiple smaller cranes to different parts of the site. Such a strategy gave the contractors the agility to move horizontally across the site, relocating the cranes to complete both interior and exterior construction.

Over a short period of time, the 47-hectare site was "rewilded" per the design team's intent, with foundations quickly being set for buildings through this process. The "valley village," or retail spine that was planned to connect the main stadia, was also connected to two underground parking structures. These garages are naturally ventilated, with the rolling landscape forming green roofs above. Additional daylight control was achieved by winged platforms Dubbeldam refers to as "solar wings," which were covered in photovoltaic arrays. The shape of these components helped guide prevalent winds down into the recessed portions of the park. The arrays in turn provided power to light the park, and specifically the retail corridor, in the evening – a time of high pedestrian volume for shopping.

STADIA

The centerpieces of the project are two stadiums that flank the ends of the sunken retail corridor. The first, the field hockey stadium, has 5,000 seats and is contained within the 125-meter free span truss. The tension ring needed for such a span is over 1 meter thick. Further blurring the boundary between building and landscape, the playing field was inset 5 meters into the park and is anchored by two large site cast concrete abutments that anchor the roof. The playing field, which the spectator seats are oriented toward, is set level, and the park banks up around it, creating a slope to nest the seating under which spectators enter through a lobby. Dubbeldam is fond of saying "nothing is flat," which supports her interest in hybrids, suggesting that the building is not just a merger of structure and landscape but of two different assembly typologies, an arena and a theatre.

The table tennis stadium is a 5,000-seat, 35,000-meter concrete structure with a curved glass façade that sets up an entry sequence and VIP areas. This hybrid concert hall/stadium takes its inspiration from history, and specifically an object called a *cong*, which is a vessel with a circular tube subtracted from its center and a square outer section. The object can be traced back some 3,400 years in ancient China and is still used in ceramics and metal

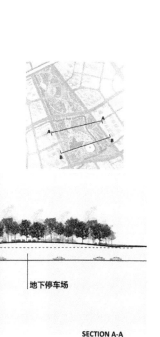

SECTION A-A

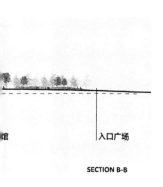

SECTION B-B

Figure 8.11 Archi-Tectonics, Asian Games, Hangzhou, China, 2022: The field hockey stadium contains a directional trussed roof spanning 125 meters. Anchored by site-cast concrete abutments, the building is set into the landscape to blur the relationship of building and site.

Figure 8.12 Archi-Tectonics, Asian Games, Hangzhou, China, 2022: Inspired by the *cong*, the architects initially studied various ways to intersect a cylinder and cube. Aware that many stadia have a direct relationship between outside and inside shapes, they intended to challenge this notion and create a geometrically hybrid building.

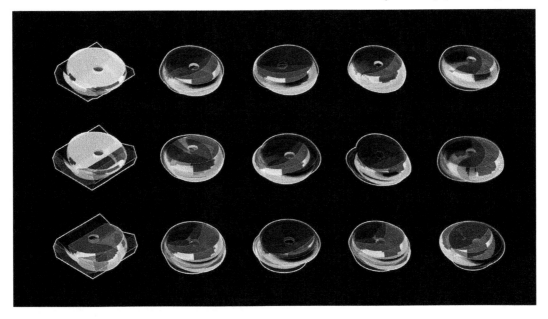

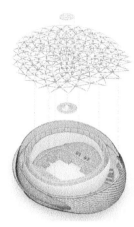

Figure 8.13 Archi-Tectonics, Asian Games, Hangzhou, China, 2022: The roof of the stadium is a large-span structural solution called a suspendome, which is widely used in sports buildings. The roof is prestressed to have much influence on the translation of internal forces.

Figure 8.14 Archi-Tectonics, Asian Games, Hangzhou, China, 2022: The 125-meter roof span of the field hockey stadium is dramatic, and covers the entire spectator seating area, which is organized on one side of the playing field. The field itself is set 5 meters into the ground to further blur the relationship of landscape and the structure, and contains large concrete abutments to anchor the large span roof.

work today. While there is some evidence to support the premise that the object had ritualistic purposes, its specific use is unknown, making it a strange hybrid, both historically and geometrically. Not unlike the cong, the building needed to be easily converted from an event space to a concert hall to a stadium, while still meeting all criteria required for the Games.

Moving away from the cylinder-cube intersection, the architects eventually began studying the asymmetrical intersection of two cylinders, which they were able to describe in terms of a radial relationship to two centers – like an ellipse – that also set up relationships between seating and stage or field. The zones that were created could be at once inside the contained volume but outside of one of the original cylinders, creating an irreducible shape that lent itself to a concert hall typology. Dubbeldam refers to the building organization as a "set of wobbly rings" that is asymmetrical and constantly negotiating inside and outside with itself – an *object with/in an object*. She contends the building is *in*stable, as opposed to being fixed as either stable or unstable, a kind of both-and condition that furthers the hybrid logic of the project.

The building's material logic is formed around the asymmetrical rings, which have a variable relationship to one another. The building's geometry was originally massed in Rhinoceros, but the intersections ultimately needed a higher level of constraint. This was furthered by Dubbeldam's insistence that the stadium needed to be a column-free space. In addition to blocking primary views

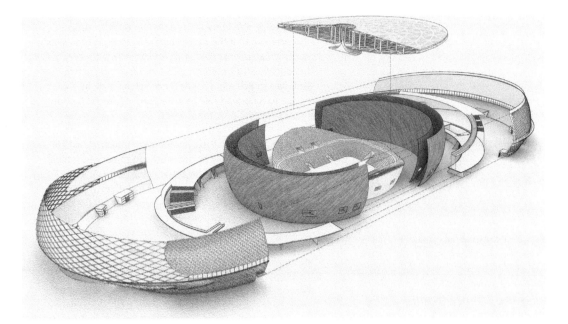

Figure 8.15 Archi-Tectonics, Asian Games, Hangzhou, China, 2022: The object with/in strategy allowed the design team to deploy a hybrid seating organization so that both stadium- and arena-type events could be held. Dubbeldam was very conscious of the structure's post-Asian Games use, so the building would not become a "white elephant," an underutilized piece of urban infrastructure in the city of Hangzhou.

Figure 8.16 Archi-Tectonics, Asian Games, Hangzhou, China, 2022: The entry sequence of the table tennis stadium dynamically expresses the relationship between the center bowl, which is clad in wood and forms a hybrid theatre/arena seating arrangement, and the eccentrically hung façade, which is clad with solar glazing and brass-finished shingles.

Figure 8.17 Archi-Tectonics, Asian Games, Hangzhou, China, 2022: By extruding a variable (triangular) frame for the façade glazing system of the table tennis stadium, the design team was able to use planar glass panels; no curvature was needed. The scale of the panels was studied with respect to the overall building to ensure the appearance of soft curvature was achieved.

from spectator to stage/field, columns would preclude the hybrid uses imagined for the structure.

The team's structural roof solution – the suspendome – is a sort of hybrid space frame with an inner and outer ring used in the arena design that is deepest at its center and has a series of tensile members that tie the deep structure back to an inner bowl, which is circular and gives form to the stadium seating. The outer ring is cantilevered

and elliptical in shape and provides support for an exterior façade, which is hung. Dubbeldam calls it a "big mushroom suspended from above." This solution allowed for a seating hybrid as well, so that both area and stadium seating organizations could be utilized, while still meeting all competition requirements set forth.

The suspendome features a massive skylight for daylighting as well as passive ventilation, drawing hotter air up and out of the structure. The edges of the structure also have smaller horizontal windows, controlled by temperature sensors, allowing them to open during periods of high heat and humidity while providing indirect natural lighting.

A large diagrid facade, hung from the roof, wraps the building at its most eccentric point. This façade is clad in glass panels, organized into 5 × 7 meter sections. Given cost-saving efforts, the architects found ways to achieve the overall curvature of the façade with flat materials. More specifically, the design team conceived a system of tapered eyelets created by the intersection of aluminum frames that would allow for a variable nesting surface for glazing panels. The system followed the overall curvature of the roof above and was fit with planar glass panels. The costs saved by not curving glass went into the performance of the panels themselves – double glazed panels with a solar film were used and still translated into a savings of several million dollars.

The building is so large, at 35,000 square meters, that the 5 × 7 meter proportion for the eyelets is appropriate and allows the planar glazing panels to follow the curvature of the roof, giving the appearance of a subtle bending even though the panels are flat. The building seems delicate in this way. This is gently contrasted by a series of intersecting rings clad in brass shingles laid in a diagonal pattern. The shingles, which number some 8,000 units, were also optimized through a BIM routine, allowing there to be only 85 unique shapes.

The workflow arrived at using BIM led to a series of novel optimizations. First, the roof structures were made incredibly light. At the beginning of the 2020s, the global availability of steel was low and demand was high, making its cost considerable. Using as small a volume as possible while still achieving the desired spans was a challenge. The reduction of structural volume brought other efficiencies as well, specifically, modularity. The roof of the suspendome was imagined as six discrete structural pieces that were fabricated off site and craned into place.

Dubbeldam compares the effect of the roof to a sundial sitting on the inner bowl that houses spectator seating, and delicately carries the façade that reinforces the

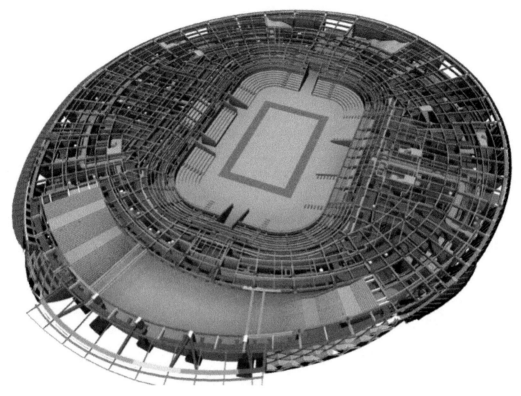

Figure 8.18 Archi-Tectonics, Asian Games, Hangzhou, China, 2022: Once the COVID pandemic of 2020–2022 effectively halted travel, architects looked to more sophisticated virtual means to ensure design intent. Working with a BIM specialist that mediated between the design and contracting teams, optimizations were found in the trussed roof structures housing the table tennis and field hockey stadia – the two principal buildings located within the park.

building's curvature. The shapes of the buildings are perhaps most notable at night, when their illuminated silhouettes are most prominent in the rolling landscape.

Adjacent to the buildings, and joined via an underground linkage is a sports complex with three Olympic-sized pools and training fields. The underground spaces, which have abundant sky lighting above, were all formed with cast-in-place concrete. Circulation through these spaces meanders, being coordinated with the gently rolling slopes of land and the structures nested within.

Indeed, the buildings will meet their hybrid goals; following construction but prior to the commencement of the games an event was held in the table tennis stadium for 2,500 university faculty and students.

ENDNOTES

1 Herbert Muschamp, "A Pair of Crystal Gems Right for Their Setting," *The New York Times*, January 14, 2001.
2 Winka Dubbeldam, *Strange Objects, New Solids and Massive Things*: Archi-Tectonics (Actar, 2022).

IMAGES

pp 132–135, 137, 139, 141–147, 149 © Winka Dubbeldam, ARCHI-TECTONICS

Figure 8.19 Archi-Tectonics, Asian Games, Hangzhou, China, 2022: Although the table tennis stadium is some 25 meters tall, the strategy of nesting the large structure into the rolling landscape of the park gives the project an overall feel of horizontality, with land bridges forming an intricate play with the waterway, connecting various programs within the site, including shopping, playing fields, and spectator sports, with passive ecological strategies including the placement of wetlands and bioswales.

PART 3

9 | ON CONSTRUCTION I

Everybody's building ships and boats
Some are building monuments
And some are jotting down notes . . .
— Bob Dylan
The Mighty Quinn (1967)

Aspects of mass-standardization, and efficiency schemas, born in the twentieth century through both the advancement of manufacturing technologies as well as accessibility to computing and the rise of information technology, have gradually affected the broader construction industry. While the industry has been traditionally slow to respond to change, several hybrid technology and construction companies have recently formed, with varied results.[1] Architects have embraced these developments and found ways to use them to further our scope and work; however, questions remain in terms of our ability to meaningfully interact in downstream construction operations, especially as those interactions are increasingly automated and involve some level of human–computer interaction (HCI).

MASS-PRODUCTION AND INFORMATION TECHNOLOGY

Hybridization attempts have been based on bringing end-to-end manufacturing processes to the construction industry, cutting out a series of middle steps, including suppliers, owner's representatives, and even architects, replacing them with in-house employees to streamline what has conventionally been a messy process. The idea is simple enough: by "skipping the middlemen and selling . . . directly to general contractors[2]" greater efficiencies, translating into both cost and time-savings, could be achieved. Such goals were embraced by technology startup Katerra whose goals of streamlining the construction process extended to general contracting – by buying local companies in regions they worked – and contemplating a modular production basis for their products. Issues arose, stemming from Katerra's embrace of principles of mass-customization. "Rather than mass-produce a single type of building, Katerra built offices, hotels, single-family homes and apartment buildings of varying heights. That made it much harder to mass-produce prefabricated parts in factories and reduce costs because a wall panel designed for a three-story apartment building didn't work for a 10-story building.[3]" The misalignment of customization goals with the efficiencies of standardization is one that information technologists

Figure 9.1 Katerra, Panelized Construction, 2019: Katerra, a company imagined in Silicon Valley and funded by $3 billion in investment, operated for about five years before declaring bankruptcy in June 2021. According to the *Wall Street Journal*, the company agreed to build projects in various US markets prior to formulating how to efficiently mass-produce building components that would be used in their construction projects. This inability to optimize what quickly became a large-scale operation prohibited the company's success.

suggest IT can correct, as seen in the manufacture of products such as semiconductors – Katerra's first chief executive came from the electronics manufacturing sector.

Katerra was an interesting experiment to say the least, as it leveraged information technology in the service of construction but also sought to change the practice of building itself and the stakeholders who come together under sometimes contentious circumstances within the process. The latter generally follows movements that have been in place since the 1980s, with the Toyota Production System (TPS), which led to popularization of the brand in the US, and even inspired a film starring Michael Keaton, *Gung Ho,* in 1986. More recently, the TPS has been the basis for a theory of *lean production*, a precursor to lean construction. According to Lauri Koskela, a professor of Construction and Project Management at the University of Huddersfield in the UK, lean theory is a generalization of Toyota's principles. He suggests the traditional ontological approach to production has been in the division of objects (wholes) into their constituent parts, where natural science will help locate an "explanation at the lowest possible level." Lean augments this *part-to-whole* understanding by introducing ideas of flow- and value generation-theory, with the former including both material and information flows. In this expanded understanding, what Koskela refers to as "good design information" refers to both design data itself and its embodiment in material products from the viewpoint of customer needs, representing value generation.[4]

Specifically, Lean has four key components:

1   A *just-in-time* inventory method where goods are received as they are needed, so no or minimal storage is required
2   Total quality management, a workplace climate where employees of an organization continuously expand their ability to provide products on-demand

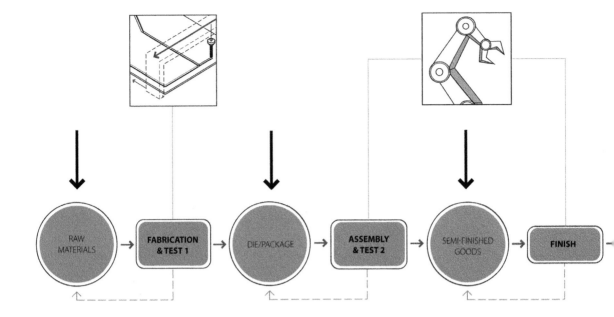

3   Total productive maintenance, a process of optimizing equipment through workplace collaboration in the production of goods
4   Human resources management

Lean in this sense shares similarities with business process reengineering (BPR), a 1990s-era management strategy for enterprises that promoted information technology in the design of workflows and business processes within a company. The move, as referenced by Alec Sharp and Patrick McDermott, as well as others, sought to reinterpret ideas of quality in products by understanding business as an interrelated system heavily reliant on computing.[5]

LEAN CONSTRUCTION

It should be noted that Lean as a theory of production and Lean Construction, while sharing similar goals of optimization, are quite different. Lean as a process can be applied to an organization, or specific parts of one, with impact to both products and their production. Construction, on the other hand, tends to be a bespoke and one-off process, on specific sites within specific contexts producing buildings of various organization, typology, and material assembly. Construction timescales,

Figure 9.2 GRO Architects, Manufacturing Process after Daniel Rivera, 2022: A simplified end-to-end manufacturing processes from the semi-conductor industry. While some "external control" is required, it is primarily in procurement of goods. The flow of materials, from left to right, follows a feedback loop of "local control," which keeps *all* of the production within the factory environment. In some cases, this model has been ported to the construction industry.

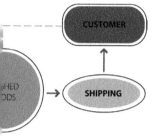

though longer than design timescales for the buildings produced, are also much shorter than the lifespans of most organizations.

Still the development of lean processes, both in and out of the construction industry tend to track with the growth of building information modeling systems in the design and construction of buildings. Both are routed in a robust utilization of information technology. By applying the concepts of flow and value to a design and construction process, especially one already BIM-centric, there emerge ideas about quality in both the object (building) itself, and the processes that allowed its actualization. Bhargav Dave, a researcher at Aalto University, and Rafael Sacks, a professor of Construction Engineering and Management at the Technion-Israel Institute of Technology, suggest that the emergence of BIM and Lean principles is not one of coincidence, with the two closing the gap between product and process. Though they suggest such a confluence can greatly augment functionality in construction management, the specific processes they locate are already available in many BIM software packages. These include the integration of various discipline models, clash detection, quantity takeoffs and bills of materials, and the visual tracking of construction progress within the model environment[6].

The codification of specific tools utilized in the design and delivery of buildings, combined with a higher-level approach to managing the process of building would obviously lead to further efficiencies and optimizations of a rather messy process – construction; however, the application of a top-down set of management practices may not be realistic for the construction site. It should also be noted that the buildings architects design will not all be built by the same general contractor, and the buildings constructed by a general contractor will not all be designed by the same architect. Familiarity and personal relationships therefore will not always be applicable in a construction process. For Bo Terje Kalsaas, a professor of Engineering Science as the University of Agder, mutual "interdependency drives iteration in design, as the conversation between interdependent specialists must go through an indeterminate number of cycles to achieve alignment."[7] In this case, BIM certainly allows for further iteration, and communication, with those tasked with both design and implementation.

MEANS AND METHODS IN CONSTRUCTION

The flipside of such thinking brings us back to the Katerra model, in which all interactions are *controlled* within a single entity or organization to decrease the siloed nature of traditional owner-architect-contractor interactions. The division of labor and responsibilities still today is of very specific and contractual circumstance, with the

| ALIGNMENT OF RESOURCES | COORDINATION | JUST-IN-TIME RESOURCE PULL |

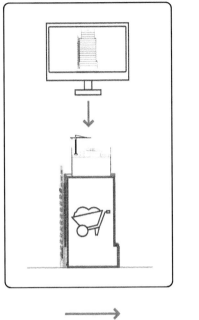 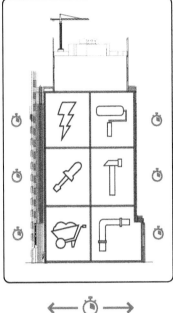 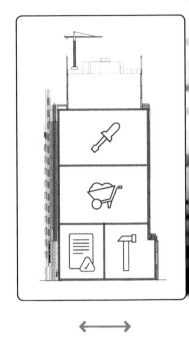

BIM/DATA MODELING **REAL-TIME COLLABORATION (SUB-CONTACTORS)** **FORWARD PLANNING**

design intent remaining the intellectual property of the architect and design team and its implementation being the specific scope of the general contractor and their subcontractors. The techniques and processes utilized by the contracting team to this end, termed *means and methods* have traditionally been a line in the figurative sand that the design team does not cross, primarily because the design team is only tasked with creating the design intent and not implementing it. The contractual limitations of this circumstance are also related to the types and levels of insurance carried by the various entities entrusted with creating a building, with architects regularly carrying "errors and omissions" (EO) insurance. EO is a form of professional liability insurance that covers errors or misrepresentations in the contract documents put forth by the design team, but specifically does not cover construction errors. These would be addressed by insurance policies held by the contracting team. This is why many architects simply "observe" site activities – standard operating procedures preclude a more meaningful involvement.

Common building issues such as water penetration and mold proliferation, heat loss due to improperly sealed envelopes, as well as effects of water on the exterior of buildings, including performance lessening of cladding materials, are products of local conditions including weather and craft. These are intensified by climate

Figure 9.3 GRO Architects, Lean Processes after the World Economic Forum Future of Construction project, 2023: Lean construction methods promise an integration of BIM with Lean production theory and suggest optimizations within the management of construction projects. While many of these process-based operations are already commercially available in BIM software packages, their application occurs through the management of a project site.[8]

CONTINUOUS IMPROVEMENT

EXECUTION AND OPERATION
(DIGITAL TWIN)

change and are rarely found to be singular in error. In a time when the expanded scope of the architect increasingly allows for interactions with machines and manufacturing schemas, how will contractual requirements change and how willing will architects be to take on increased liability as our work involves manufacturing and material assemblies?

LARGER APPLICATION OF 3D PRINTING

Three-dimensional printing technologies, an additive material process where material is distributed just like a printer distributes ink, building up a three-dimensional shape, have been increasingly accessible to architects since the end of the twentieth century, when stereolithography machines were used for model-making and prototyping. The technology became commonplace in architecture schools several years later, and today desktop models that print with ABS (acrylonitrile butadiene styrene) plastic or PLA (polylactic acid), a material derived from corn, are available for about $2,000. The printing material is extruded to a filament that is fed from a wheel to the printing tool. These 3D models are easily workable and can be painted and glued together. 3D printers provide the benefit of cost-effective translation to allow designers to quickly move between virtual and actual models.

It would only be a matter of time until this technology was scaled, and with the proliferation of robot arms, which have also become economically accessible, many architecture schools have adopted them as multi-axis additive and subtractive tools for model-making. While traditional flatbed CNC routers could only cut in the conventional Cartesian directions of X, Y, and Z, and were prone to errors involving undercutting, robot arms have joints that allow them to spin in many directions and angles, allowing for greater flexibility in subtractive modeling. The arms can also accept a number of end-tools used for a wider range of tooling operations and have a greater capacity for customization through programming.

Figure 9.4 GRO Architects, 3D Printing, Jersey City, NJ, 2022: Desktop 3D printers have been accessible to architects and designers since the turn of the century and have revolutionized virtual-to-actual design operations, allowing for a seamless transition from digital to physical objects and environments.

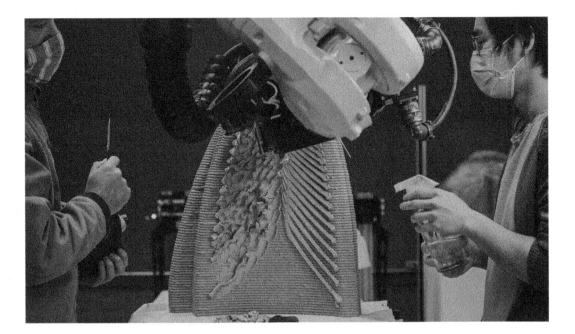

As the scale of 3D printed parts increased so did their application, with an obvious one being the creation of custom form liners for precast concrete panels. First used in the late 1960s, form liners have been applied to add a finish to the exterior of a panel and traditional uses include the addition of brick or stone veneers to architectural or structural panels. Form liners had been manufactured with polyurethane for its reusability, but ABS plastic has become an industry standard. The use of 3D printers for form liner production allows limitless customization of the sheet, made with a material already common to the industry.

Outfitting robot arms with extruding nozzles allows for additive modeling. This is now being explored to deposit cementitious material at full or near full scales. At the time of writing, multiple companies have launched promising full-scale concrete printing services within a direct-to-manufacture interface. The premise is similar to using a robot arm to deposit a viscous material that cures but utilizes an overhead industrial gantry or similar mechanized system to move in larger-dimensional increments across Cartesian space and a larger vat from which the base material is distributed. Companies such as MudBots, located in Midvale, Utah, offer in-house solutions but also sell manufactured printing solutions and even franchising opportunities for 3D concrete printers. MudBots suggests that using its products can reduce production times by as much as 70 percent,[9] as traditional formwork becomes unnecessary. Thus, its production methods are more cost-effective and less wasteful. Reusability of formwork – or in this case, its elimination – is an important aspect of material consideration. The company produces machines of various sizes from

Figure 9.5 Weitzman School of Design, University of Pennsylvania, 2022: The Robotic and Additive Manufacturing Lab at the Department of Architecture at Penn is exploring extruding nozzles to extend the use of robot arms as additive printers. The arms deliver a viscous cementitious material that, while still wet, can hold form while curing.

6' × 6' × 5' high to 100' × 50' × 10' high; however, existing literature suggests only smaller machines have to date been utilized in production.

TECHNIQUE > PROCESS > WORKFLOW

The wider interest in how the work of the architect is more specifically aligned with construction practices has to do with modeling itself, or the adoption of new models, that is, modeled objects as well as modeled processes. This in itself is an expansion of how architect's *work*, and there has been a lineage of modeling terminology that loosely follows Andrew Witt's computational timeline, which is presented in the *On Practice* chapter of this book. As designers became more versed in rapid prototyping and CNC machining of models and parts, design operations are needed to encompass techniques to allow model geometry to be managed for export to a CNC interface. Techniques for closing geometric solids, unfolding or unrolling surface geometry, or nesting parts on a virtual "sheet" of material, first manually and then through automated routines, became commonplace in the late-1990s, and later part of an expanding architectural design process. In turn, that process widened to include concepts of assembly and specific ideas about how things were made. A process of geometric rationalization through a series of both digital and analog techniques was originally posed in *BIM Design* in 2014 in which more elemental geometric objects – lines and curves – were developed into more complex ones – surfaces – and to solid assemblies. At the time, such an actualization was being studied by architects on smaller-scale projects and exhibitions, where scope and associated liability was low, though some

Figure 9.6 UNStudio, Frankfurt Four, Frankfurt, DL; Design for Assembly, 2017: UNStudio in collaboration with Digitales Bauen are exploring modular construction in large-scale formats. In Frankfurt Four, a 300,000 m2 project spanning four new buildings, the process involves breaking down a building into a series of modules and applying rule-based modular concepts for off-site assembly, transmitted to all members of the design and construction team. While such a process is not "BIM specific" the flexibility of modeling platforms to allow designers to understand components and part-to-whole assemblies facilitates novel workflows and delivery. Such work is often driven by mechanical, electrical, and plumbing (MEP) constraints, which conventionally defy modular logics.

BIM + Modular Building

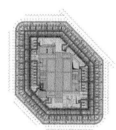
Integrationsplan Technik Decke Turm 1

Integrationsplan Technik Decke Lüftung Turm 1

architects were already looking to larger-scale building transformations. The term *workflow* was introduced shortly thereafter. Defined as a progression of steps (tasks, events, interactions) that comprise a work process, involve two or more persons, and create or add value to (an) organization's activities,[10] workflows pertained to larger-scale construction operations and further aligned the work of the architect with more downstream construction procedures.

These advancements, rooted in a geometric basis that connected our digital design work to various types

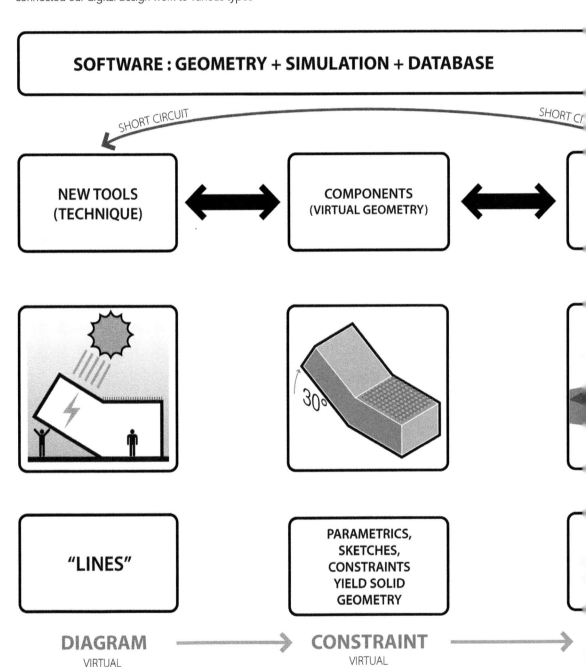

of physical output, have given way to more complex workflows that consider multiple site activities and include processes as well as spaces and components. Still, it should be acknowledged that the messiness of construction involves the complexities of personnel (humans) and their interactions with others (both human and non-human) in the execution of a building. As we continue toward a more automated basis of production, these interactions will be rethought, if not lessened, and the construction process will become leaner not so much by management but by design.

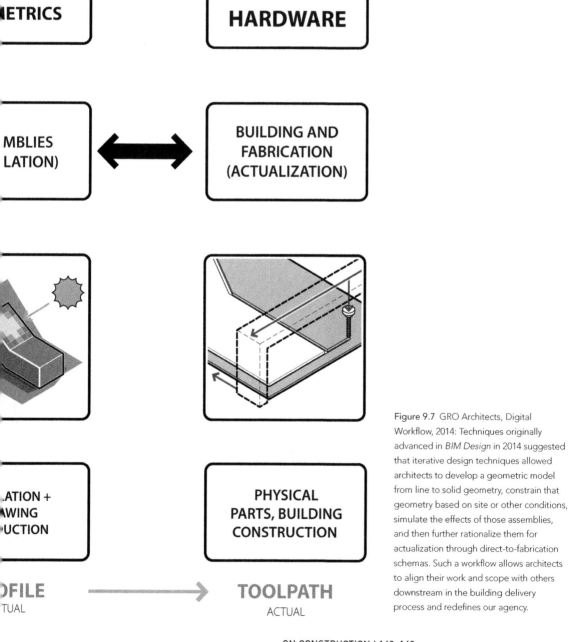

Figure 9.7 GRO Architects, Digital Workflow, 2014: Techniques originally advanced in *BIM Design* in 2014 suggested that iterative design techniques allowed architects to develop a geometric model from line to solid geometry, constrain that geometry based on site or other conditions, simulate the effects of those assemblies, and then further rationalize them for actualization through direct-to-fabrication schemas. Such a workflow allows architects to align their work and scope with others downstream in the building delivery process and redefines our agency.

NOTES

1 Konrad Putzier and Eliot Brown, "How a SoftBank-Backed Construction Startup Burned Through $3 Billion," *Wall Street Journal,* June 29, 2021, https://www.wsj.com/articles/how-a-softbank-backed-construction-startup-burned-through-3-billion-11624959000 accessed 12 May 2022.
2 Ibid.
3 Ibid.
4 Patricia Tzortzopoulos, et al. *Lean Construction Core Concepts and New Frontiers* (Taylor & Francis, 2020).
5 Alec Sharp and Patrick McDermott, "A Brief History – How the Enterprise Came to Be Process Oriented," *Workflow Modeling: Tools for Process Improvement and Application Development* (Boston, MA: Artech House), 2008.
6 Tzortzopoulos.
7 Tzortzopoulos, p. 214.
8 World Economc Forum, "Shaping the Future of Construction: A Breakthrough in Mindset and Technology", May 2016 see https://www3.weforum.org/docs/WEF_Shaping_the_Future_of_Construction_full_report__.pdf.
9 "Going Up: From Here to There," MudBots, https://www.mudbots.com/going-up.php accessed 26 May 2022.
10 Richard Garber, ed. *Workflows: Expanding Architecture's Territory in the Design and Delivery of Buildings* (Hoboken, NJ: Wiley, 2017).

IMAGES

pp 155–159, 162–163 © GRO Architects; pp 160 © Weitzman School of Design, University of Pennsylvania; pp 161 © UNStudio

10 | COMPONIBILE, BY REMOTE CONTROL, GRO ARCHITECTS

AMATERIAL VIRTUAL ENVIRONMENTS

In late March 2020, we were looking forward to a trip to Italy – this was to be a stopover while traveling to Tel Aviv with my Penn students – with the specific intent of reviewing built-in furniture that was being manufactured for a new housing project of ours in Jersey City. We would be heading to a factory in Ancona, northeast of Rome on the Adriatic Sea. Our team had been communicating with the fabricators there for months, transmitting both drawings and raw digital data for the fabrication of these objects that would figure prominently in the interior organization of our building. The ease we felt with these transmissions, the two-way flow of data that would ultimately yield these very specific three-dimensional objects, felt very twenty-first century.

Figure 10.1 GRO Architects, Nest Micro-Housing, Jersey City, NJ, 2014–21: Density serves as the impetus for flexible and sustainable living while promoting community development in Nest Micro-Housing, a residential project located minutes from the Journal Square Transportation Center in Jersey City. The 38,000 square-foot community, containing 122 dwelling units, is a progressive housing alternative for individuals seeking privacy, modern amenities, and economy while liberating them from *stuff*. Nest's 220 square-foot units are designed to be highly efficient with flexible furnishings, where different configurations for living, dining, or sleeping, and variations in between, are possible. The project was an opportunity to imagine part-to-whole relationships at a variety of scales including component, unit, assembly, and larger cluster within the developing neighborhood.

The trip, of course, did not happen (nor did a planned family holiday to Tuscany later that year), due to the COVID-19 pandemic. What did was a months-long assessment via WhatsApp, FaceTime, and ultimately Zoom, where we awkwardly watched our counterparts at the factory in Ancona flip tabletops, unfold mattress platforms, and demonstrate the smooth operation of a variety of pulls and hinges through the window of a computer monitor – or worse, a phone. To further complicate things, the mockups we were evaluating were custom furniture designed to be installed in a micro-housing project of ours that was well under construction. The 122 dwellings that comprise the building are only 220 square feet, which at 11′ × 20′ are significantly smaller than typical studio apartments. We had comprehensively modeled the building three-dimensionally and collaborated closely with our consultants to simulate building components and their performance. After years of working virtually to bring the project to its material actualization, we thoroughly understood its physical qualities and the intricacies of the building systems that were compressed into an envelope fixed by zoning constraint but without density limitation.

The whole thing was oddly *amaterial* – as opposed to *immaterial* – as the pandemic set forth significant events in our world. We were unable to feel the sleekness of a

Figure 10.2a–c GRO Architects, Nest Micro-Housing, Jersey City, NJ, 2014–21: Micro-housing brings with it some challenges beyond more standard types of residential buildings. First, there is generally a higher degree of coordination required across the architectural, engineering, and construction team due to building density. As average one-bedroom apartment containing a bathroom and kitchen is about 750 square feet in area; there are nearly three times the amount of plumbing and gas risers, equipment, and fixtures within the same 750 square feet taken up by three micro-apartments. The design team required the mechanical consultant to model all plumbing and gas risers three-dimensionally and integrate this work with the architectural scope, tendering a digital model to ensure there were no spatial conflicts between trades and study both a process and order of construction. In compliance with construction code, the building's size defined its construction classification as Type IIIA, which required noncombustible exterior wall assemblies and interior building elements made of light gauge steel.

Figure 10.2a–c (Continued)

surface or the robustness of a joint or comprehend the relative scale of real components as they were organized and configured within a fixed envelope. It was a strange sort of hybridization of what we had hoped our practice of architecture to be, at once digitally advanced yet with a highly specific material actualization, collapsed through the medium of a screen. We wanted to embrace this *remote control* as the logical extension of the digital protocols we increasingly utilize as architects, fostering new ways to liaise and allowing expertise from a larger and more diverse set of collaborators. But the *remoteness* of the control – or lack of it – we experienced demonstrated a sort of material detachment, making our highly specific design intentions seem physically vague as they were mediated by *that* window that put our colleagues visually within feet from us, yet so far away. It reminded us that as a profession we had been here before, and that this experience had both sociocultural significance as well as pedagogical consequences.

The period leading to the widespread publication, in 1485, of Leon Battista Alberti's treatise on our nascent profession, De Re Aedificatoria (*On the Art of Building*)

Figure 10.2a–c (Continued)

saw tremendous achievements in both codification of the practice of architecture, as well as building methods that would be standardized through the process, and more recent scholarship has reconstructed the workings of late-medieval architects.[1] In his 1985 essay, "Gothic Architecture by Remote Control," Franklin Toker exposed an illustrated building "contract" from 1340.[2] While Toker's work identified a very clear written and graphic set of instructions for a building project in Siena, it also raised issues of how design standards, and architects in general, came to be. This concept of remote control is not new, and in fact, Alberti's treatise helped ensure that the knowledge and expertise we gained as architects would ultimately allow us to *leave* the project sites where we previously worked, specifically as expert masons or builders, and contemplate projects in other contexts and places with diverse configurations that required both cerebral as well as material solutions for problems of building.

Though Alberti clearly intended for this design-centric mode of practice, it is important to recall that his books are about both building and buildings, outlining improvements in technology while also codifying many standards once held in relative secrecy by the guild societies that existed prior. His books are headlined with specific typologies of the time, including sacred and mercantile buildings, as well as the proper organization of components, or liniments, that bore them. The medium of the architect, however as we know, would not be the literal material substrate of buildings but the representations that would illustrate their order and assembly – think *control*. It is no surprise that De Re Aedificatoria remains not only a present but a critical component of architectural education and pedagogy.

In the past 30 years, parts of the profession have challenged this mode of cerebral representation in favor of a more digital and materially specific way of working. Most of us at this point have experienced the satisfaction of sending a digital file to a piece of hardware such as a 3D printer or router that yielded a material translation, and many architects have successfully integrated this way of working into their larger design workflows. The promise of these operations is that they were bringing us closer to building – the material assemblages that were formerly executed by others who interpreted our work; but through the lens of 2020, we had come to understand our work, in all its simulated glory as *remote* in more complex ways.

In remoteness, relationships and inputs for design become essential in different ways, and the sequence of demonstration, or the experience of it, whether scripted or not, allowed us to see orders other than what we may have imagined. It allowed us to understand our work as an aggregation of building components – modules contained within a larger modules with specific functionalities – as opposed to a more conventionally programmed spatial sequence.

We are reminded of Wittkower's observations about the importance of the repeated module and relationship of part to whole that was so fundamental to the spatial organization of humanist architecture. This was exemplified in his analysis of Palladio's villas in which the rules of humanist proportion and ratio previously reserved for the design of harmonious facades was used to establish a spatial 3D logic of the entire villa. In the Palladian villa, the module of the column informed the dimensions of a room and its relationship to other rooms, in correspondence with the building as a whole. The humanist tradition was prevalent at a smaller scale in the evolution of Italian furniture from early utilitarian designs to the highly decorative furnishings of the Renaissance, where proportional relationships between components were set in relation to ideal Vitruvian ratios. It was in the Industrial Revolution that Italian furniture design ultimately made its full embrace of the module with the emergence of mass production.[3]

ASSEMBLIES WITHIN ASSEMBLIES

Nest is located within a progressive redevelopment area in the Journal Square neighborhood of Jersey City, that encourages high density, and therefore sustainability, by supporting a concentration of resources equitably and efficiently distributed to a larger user base within a relatively small urban footprint. Journal Square hosts a regional transit hub with buses, taxis, and the PATH train, which travels east to the World Trade Center and Penn Station in Manhattan, and to Newark Penn Station to the west, with Air Train connectivity to Newark Liberty Airport. The redevelopment plan largely controlled the building's bulk; however, the site offered unique design opportunities. Permitted building height is six stories, with 100 percent coverage on the ground floor and the upper floors controlled by a 30-foot rear-yard setback. The site slopes over 10 feet from the rear-yard to the sidewalk and permitted a "walk-out basement" at sidewalk level, with the ground floor, which is the first residential floor, at near 100 percent coverage, above. The perimeter condition of the basement level was coordinated through geotechnical exploration for minimal excavation of rock. That floor is accessed directly from Academy Street, at the building's right-of-way, and is programmed as a café that is open to the public, with residents having after-hours access for co-working.

Modular furniture components, now produced in repetition as a set of parts that could be joined to form wholes reached another level of innovation in the 1960s, as dynamic assemblies of modular components were set in such "perfect" geometric relationship to one another that the various parts were designed not only to move but *transform*. The Snaidero "Spazio Vivo" modular kitchen designed by Virgilio Forchiassin in 1968, in the permanent

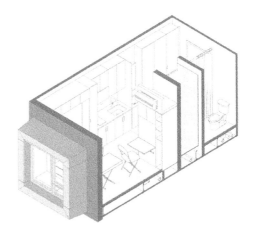

Figure 10.3 GRO Architects, Nest Micro-Housing, Jersey City, NJ, 2014–21: The basic unit is designed for a single occupancy with standard size of 11′-0″ × 20′-0′ deep. The module is subdivided into a "wet zone" and a "dry zone." The wet zone, which is positioned at the unit's interior, contains requisite bathroom fixtures including shower, toilet, and vanity; organized individually within the space, with accessibility constraints being measured to fixture as opposed to grouping them into a bathroom. Occupants enter through the wet zone, which has individual stalls dedicated for the toilet and shower, grouped on one wall. The sink and storage closets line the opposite wall.

Figure 10.5 GRO Architects, Nest Micro-Housing, Jersey City, NJ, 2014–21: The building's permitted height is six stories, with 100 percent coverage on the ground floor and the upper floors controlled by a 30 feet rear-yard setback. Given the sloped site, dimensional measurements from grade allowed the ground floor to actually be raised up one level, where units flank the perimeter of the site. In the center, amenity spaces, including a gym and media room – spaces that do not programmatically require natural light – are inserted.

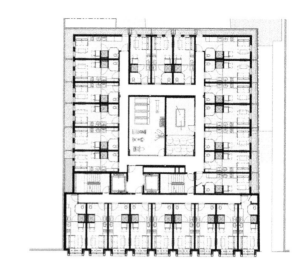

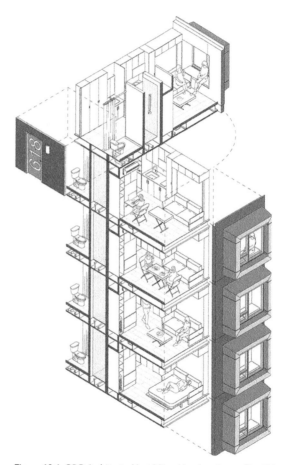

Figure 10.4 GRO Architects, Nest Micro-Housing, Jersey City, NJ, 2014–21: The dry zone contains a kitchen area with sink and refrigerator as well as a two-burner ceramic cooktop with a convection microwave mounted above and a living area with a hybrid couch/bed for sleeping, ample storage, and modest fold-down work surface. The design team worked intensively with an Italian millwork fabricator on a series of space-saving furniture pieces that were installed in situ, so all units are occupied as furnished, the concept being that occupants need only to arrive with their clothing and a laptop.

Figure 10.6 GRO Architects, Nest Micro-Housing, Jersey City, NJ, 2014–21: The micro-units have moving components that allow for spatial transformation. The "dry zone" where living occurs has at least four discreet configurations that allow for sleeping, sitting on the couch or in a window box, dining, and working from a foldable worksurface, all occurring within the same 10' x 12' (120 square feet) space and transformable by a single human.

collection at the Museum of Modern Art (MoMA), was especially innovative for creating a single "island" component comprising a series of moving blocks that were hinged to the base units and could open and close according to their function. Mechanized components became increasingly popular for their capacity to transform not only themselves, but the space they occupied, making the domestic work of living a highly efficient experience that allowed for presumably more space and time for leisure.

At Nest, the built-in cabinetry forms the basic module that is repeated to create storage opportunities wherever possible, or to accommodate compact appliances needed to prepare a full meal for friends that will join around the pop-up dining table with folding chairs that slide behind the refrigerator cabinet. The cabinetry with its hinges and hardware orchestrates various iterations of use that organize the living space within each larger unit module, which is then arrayed on each floor and stacked. It was the proportion, part-to-whole relationships, and the transformation of space achieved through the module that made the cabinetry such a critical aspect of the design, and why the tradition of *componibile* in Italy was so essential to our conception of this project.

The tradition of craftsmanship in Italian manufacturing is well known. We, like many architects and interior designers, have specified Italian cabinetry for custom residential design projects given their high level of quality and craft. But it was on a tour through various furniture factories years earlier that had exposed us to the high level of material expertise and craft especially in the region of Brianza, located 10 miles north of Milan. This region became the center of Italian furniture production in the 1770s due to the grand visions of the Archduke Ferdinand of Austria-Este, who sought to furnish his vast palace, the Villa Reale in the town of Monza.[4]

The sheer number of furnishings required was so great that he invited artisans from all over Europe to come to Brianza to work for him. Once the palace interiors were complete, the craftsman stayed, establishing the region as a center for furniture production, which remains to this day. Crafts-work and an inherent understanding of proportional relationships between movable components are linked to one another by the requirement that movable parts be especially well-made to allow for a smooth and robust operation. The transformative capacity of a living unit to morph from a dining space to a living space, or from a work space to a sleeping space, with ease was the basis for the design of the micro-units, which though quite small at 220 square feet could be conceptually expanded at least threefold through modular logics. With these semiautonomous components, highly specific intentions could seem more ad hoc, and gave us insight into various

Figure 10.7 GRO Architects, Nest Micro-Housing, Jersey City, NJ, 2014–21: Nest privileges a sustainable lifestyle by providing fully furnished dwelling units allowing for occupancy with not much more than oneself. Furniture was produced specifically for the project and supports local reconfiguration that allows for a variety of living scenarios.

ways one might occupy our planned spaces instead of our perceived or expected manner – again think *control*.

IN REMOTENESS. . .

Further complicating the remoteness of our situation was the disfunction of global material supply chains, where the previous (and relative) ease of sourcing both raw materials and specific tools and parts for assembly, were substantially disrupted by the global pandemic through closed borders, diminished workforces, and our limited access to them. Tapping these limited systems became part of our architectural purview in remoteness. There was a moment in April 2020 when Italy was feeling the brunt of the pandemic that we started to explore other sources for the furnishings as production at the factory had come to a halt. But the pandemic was volatile and unpredictable – and the tradition of craftmanship and the innate understanding of modular thinking at the basis of Italian design and production gave us back some sense of control.

Figure 10.8 GRO Architects, Nest Micro-Housing, Jersey City, NJ, 2014–21: Each dwelling unit facing the right-of-way contains a glass window-wall with both fixed and operable glazing that is coordinated with prevalent sun angles to reduce the amount of active heating and cooling. Given the compact size of the units, condensing units for conditioned air are shared between three to four units, while being subzoned within each unit.

Precision, the highly specific ways joints came together, the overlaps between parts, and the ways we perceived component coloration through the screen became both more important and apparent. The ways these components interacted became more important than the whole – the dwelling unit itself. The need to remotely convey these component assemblies led us to process our digital data to create new visualizations for those tasked with implementing them. From our design models we studied new ways of understanding poché. In modern building techniques most wall assemblies provide for a void space, an interstitial space that, while not inhabited by humans, is essential in that it is occupied by the systems that support them. The heightened need for precision to resolve potential material conflicts was intensified in this project given the density of dwelling units, and in turn the density of building systems. Building code bounds the smallest allowable domestic space, and our unit design tested those limits through a series of diagrams that sought to define "living space." The material density of structure, plumbing, mechanical systems and living space, packed into a constrained zoning envelope truly sets the limits for this project. There is minimal interstitial/in-between/unused space that can be abstracted into poché.

The irony of designing flexible furnishings for a micro-housing project during a global pandemic which forced us to stay inside and work from home was not lost on us or our design team. As we sought to transform our own dining table into a desk or bedroom into a studio while our children went to school in our living room, it

Figure 10.9 GRO Architects, Nest Micro-Housing, Jersey City, NJ, 2014–21: The front façade, which is south facing, contains a series of 45 window boxes. In addition to extending interior seating space, the window boxes project into the right-of-way 21", determined as optimal for solar shading in the summer while absorbing winter heat energy.

Figure 10.10 GRO Architects, Nest Micro-Housing, Jersey City, NJ, 2014–21: With unit demising walls on 12'-0" centers, the building functions as a high-mass assembly with sound attenuation and performance requirements making the structure highly insulative, allowing the building to passively retain indoor air temperatures.

became clear that the pandemic mandated a level of transformation and flexibility within our own spaces. If there was already an interest in transformable furnishings prior to 2020, the new nature of remoteness accelerated and compounded it, especially within the urban context.

Toker observes that the arrival of architecture by remote control in Italy was a liberating factor, not only allowing architects to build in other contexts but to experience those contexts and the cultures that gave them their form. In 2020, the pandemic compounded this notion of remoteness, requiring us to find new ways to assert our design intent while finding other ways to collaborate on a building goal. The fact that much of this collaboration occurred with the factory in Ancona reminded us of the lineage of histories that are so specific to Italy and important to our collective work as architects, and the collective origins of our profession.[5]

Figure 10.11 GRO Architects, Nest Micro-Housing, Jersey City, NJ, 2014–21: Given the small unit size and narrowness of the dwelling units through the entry and "wet zone," the design team ensured the wall assembly was both as thin as possible and had good insulative capacity, given the number of occupants in the building. Based on this objective, a decision was made to use prefabricated light-gauge wall panels. Panels were trucked to the site and craned into place. While the project was therefore not modular, there was a time- and cost-savings utilized by offsite panel fabrication.

NOTES

1 Richard Garber, "Alberti's Paradigm," *Architectural Design* 79 (2009): 88–93. https://doi.org/10.1002/ad.859
2 Franklin Toker, "Gothic Architecture by Remote Control: An Illustrated Building Contract of 1340," *The Art Bulletin* 67 (1) (March 1985): pp. 67–95.
3 Silvia Barisione, "Craftsmanship and Industrial Production of Italian Furniture During the Interwar Period," *The Journal of Modern Craft* 13 (3) (2020), 271–290, DOI: 10.1080/17496772.2020.1843781 10.1080/17496772.2020.1843781
4 Nancy Hass, "The Dynasties," *The New York Times*, April 13, 2020, https://www.nytimes.com/interactive/2020/04/13/t-magazine/italian-fashion-design-houses.html, accessed October 22, 2022.
5 I would like to thank Nicole Robertson for her contributions to and review of this text.

IMAGES
pp 166–170, 172–173, 175–177 © GRO Architects

11 | ON CONSTRUCTION II

If we as a profession are ever going to find our way to a seat of power, even the most design-driven architect must realize that construction, financing, marketing - all these things - directly or indirectly affect form, and the sooner one comes to understand these subjects the more control one will exercise[1].

— Carl Koch (1994)

With the interest in off-site fabrication and assembly increasing, and gains afforded in modular technology and quality, opportunities for architects to engage in these processes will continue to increase. Architects willing to explore more expansive workflows, and collaborations with manufacturers will not only yield more specific translations of design-intent but also expand agency. There are multiple popular prefabricated construction systems increasingly being utilized within the industry. Those being studied by architects include panelized light-gauge framing, wood modular assembly, and panelized precast concrete.

OFF-SITE AND MODULAR BUILDING

The idea of efficiency that emerged from the mass-standardization practices of the twentieth century was enabled by the re-appropriation of manufacturing efforts following the end of the Second World War. During this time, factories given to the war effort began to produce *things* for consumption, aided by the new connectivity afforded by the interstate road system. Interest at this time grew in the standardized production of construction materials and products. These included precast concrete, a newly codified use of reinforced concrete arrived at through advances in chemistry and material science that yielded panelized wall and floor products produced in climate-controlled environments and shipped to construction sites. Precast continues to be a popular, and practical, material choice for buildings such as parking garages and industrial and warehouse spaces.

Modular construction also became popularized at this time and is currently enjoying a resurgence in interest, partly due to Lean considerations given to its production. Timber remains the primary construction material for modular building, with dimensional and engineered lumber products being

Figure 11.1 GRO Architects, Nest Micro-Housing Jersey City, NJ, panelization, 2021: Several modular systems were considered for Nest Micro-Housing, a 122-unit micro apartment building completed in 2021 by GRO Architects. Given overhead wires and a public bus route on the street, use of a large crane to set modular, or precast, boxes was prohibitive. The architects opted for a light-gauge steel panelized system in which steel studs of various gauges were assembled into panels in a controlled environment and shipped to the site. A small lift was then able to set the panels floor by floor.

utilized. Light-gauge steel is also used, and allows for taller buildings, however, light-gauge framing is more commonly used in the production of off-site panels that can be more easily shipped to construction sites than modular boxes, and can be tilted into place.

MODULAR MATERIALS AND WORKFLOW

Many modular companies favor wood assembly systems and participate in the production of smaller scale structures, including one- and two-family houses, as well as multifamily housing and dormitories. Wood remains a popular and inexpensive building material for the housing market, and while heavy timber continues to find novel use in the construction of tall buildings[2], some 90 percent of new homes built in the US in 2019 were wood-framed[3]. Manufactured wood-modular systems are similar to site-built wood frame structures in that they are primarily made of dimensional lumber, and some engineered wood products such as I-joists and open-frame wood trusses for floor systems. Engineered wood products share sizing with dimensional lumber but are stronger and lighter, allowing them to span greater distances. Many jurisdictions in the US allow for a maximum of four to five stories of wood. Wood construction systems can be hybridized with up to

Figure 11.2 GRO Architects, Moxie, Jersey City, NJ, modular erection, 2021: A modular dwelling unit is picked from a flatbed truck and stacked into place. In many instances, the stresses of shipping and lifting a modular unit can be higher than the live and dead loads they will encounter during intended use.

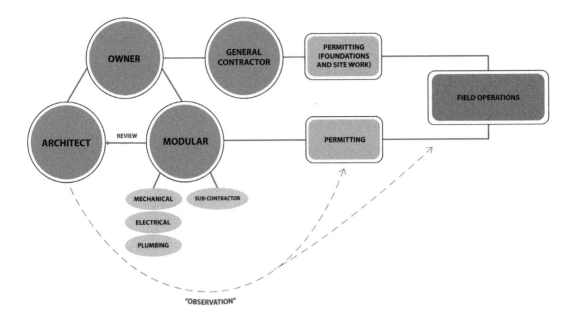

two stories of noncombustible construction materials – steel, concrete, or masonry – and can reach a maximum height of six stories and 70 feet.

In modular construction, a series of framed surfaces combine to form six-sided modular units, which are fit out in a factory with wall and floor coverings, as well as all piping, ducting, and conduit required to finish the unit. Many modular construction companies are set up to be turnkey solutions, following rules of standardization they employ "in-house" architects and engineers for design-side operations and utilize industrial facilities for the storage and assembly of components into modular units. In these instances, a "design architect" will typically perform work through schematic design and design development before handing off design intent to a modular company for further implementation. Drawing sets are transmitted to the modular builder so that an assembly logic consistent with their system can be applied. The review of the modular intent would then be tasked back to the design architect, who would review the modular assembly drawings and specifications much like the review of a shop drawing submittal.

The issue with such a workflow, especially when more novel organizations are desired, is the modular companies may not have the capacity to capture non- or less-standard design intent. This is not to suggest that modular companies are not able to *build* such solutions, but that deviation from standardized production routine(s) is inconsistent with the protocols they utilize, which are more consistent with Fordist principles than those of mass-customization we are increasingly familiar with today. Such is an opportunity for architects to both expand scope and find novel ways of integrating intent with a prefabricated construction system through information exchange with

Figure 11.3 GRO Architects, Traditional Modular Workflow, 2022: In a traditional modular workflow, architectural impact, already limited in conventional construction schemas, is further marginalized given the increase in scope undertaken by the modular fabricator. These operations can include construction documentation, design and coordination with engineering consultants including mechanical, electrical, and plumbing (MEP), and in-factory inspections and permitting.

Figure 11.4 GRO Architects, Moxie, Jersey City, NJ, shipping of modular units, 2021: Modular units are typically constrained by transportation logistics, with final delivery usually being by flat-bed truck. Units are generally held to heights of 11'-0" and widths of about 16'-0" for standard transit.

the modular manufacturer, many of which are adopting more three-dimensional workflows. Documentation then would be concerned not only with the location, dimension, and elevation of architectural spaces and components, as in a conventional drawing set, but also include a logic of assembly. Such means and methods have generally been avoided in the traditional contractual relationships between owner and architect, and owner and builder.

RESISTANCE AND ASSEMBLY

The primary difference between wood-framed and light-gauge modular systems is the structural, and therefore height, capacity and the additional need for shear members within the system. In many instances, in both wood and steel framing, stair and elevator cores used for circulation can provide some level of lateral resistance. Light-gauge floor assemblies also usually require a concrete topping, which can be applied in the factory, the field, or both to further lock together the modular units.

As modular units are shipped as completely assembled as possible, they are typically shipped in a six-sided configuration, with two side walls, a front and rear wall, and floor and ceiling. Otherwise construction of this "box" is not so different from traditional wood- or steel-frame assembly. In addition to being able to contain all fixtures and equipment, this affords structural stability for transportation and craning, for which more structural resistance may be required than for their ultimate, resting use. Factory assembly also means that the typical depth of wall and floor-ceiling assemblies will be larger than site-built configurations due to the stacking of units, many times doubling. This has implications for height with respect to zoning, as modular buildings can be 12" to 18" taller per floor – as well as in greater depth of horizontal material assemblies, which translates to less usable interior space across a site of fixed width.

Figure 11.5 GRO Architects, Moxie, Jersey City, NJ, modular building configurations, 2022: As stacked modular units have thicker wall and floor assemblies, building heights tend to be greater and the amount of interior space across a site will be lessened. Still, many see this as a modest disadvantage given the cost- and time-savings modular systems promise.

These spatial and dimensional inefficiencies are usually tolerated given the efficiencies modular assemblies afford in terms of material and assembly cost and time savings. Economies of scale are found in standardization of materials and components across multiple projects, as well as assembly taking place within a controlled interior environment. A modular production operation may also benefit from scale, if a factory has orders for multiple projects at once, it can more competitively purchase construction materials for those projects, as opposed to only buying for a single project at a time. Factories may then have better stock of the materials needed for modular fabrication.

Since modular units are shipped to project sites and erected with a crane, there is a perception that modular construction takes less time, and that translates into shorter construction times. This means shorter durations for contracting teams to be mobilized on construction sites, for the rental of construction equipment, and potentially for lower carrying costs on the sites themselves. While the perception of speed is certainly achieved – assuming a well-organized and staged project site, a modular building can go together in a matter of days – the modular units still need to be constructed in the factory, and modular plants generally have the same access to construction equipment and materials as site-construction concerns, meaning delays due to supply chain and shipping issues, such as those felt through the global pandemic of 2020–2022, can affect production and schedules. Shortages in the labor force, another phenomenon of the COVID-19 pandemic, can also bring production challenges; however, the controlled factory environment of modular production may be seen as more attractive to construction workers in regions with varied climates.

As various trades are required to fit out a modular unit – boxes include all of the plumbing, mechanical ducting, and electrical conduit required – modular companies have "in-house" labor or use third parties for all mechanical, electrical, and plumbing (MEP) roughing, and fixture installation, required. In keeping with the turnkey logic of many modular companies, it is practical to have such trades on payroll, however, since there remains a small portion of MEP scope in the field, including connections between units and connection to municipal infrastructure including domestic water and sewer and the local power grid, many companies choose to hire third-party subcontractors that are able to work both in the factory and in the field. This allows the modular company to function as a general contractor, as opposed to a subcontractor. Still, many companies will require that a local team take responsibility for the footing and foundation system, which will be site-constructed. There is a subset of smaller-scale modular contractors that use precast concrete wall systems for the foundation. These panels, Superior Walls® or similar, are slightly widened at the base, and rest atop a compacted gravel bed removing the need for a conventional footing.

Once set in the field, finish work will be required to complete the modular assembly. Interior materials are mostly applied in the factory but may need to be finished where two units come together or patched due to cracking that can occur during transit. Any specialty fixtures or requirements will also generally be installed in the field. Many companies are experimenting with the application of façade materials in the factory as well as glazing, insulation, and vapor barriers are added to exterior walls, but there remains some finish work, especially where units come together, required in the field.

Standardization routines also govern the selection of finishes and equipment offered to customers of modular buildings. This has more to do with the efficiency of having a small pool of established vendors to buy products from than the technical ability of a fabricator to customize its workflow. As such, the catalogs of products available can be limited, which can be a restriction to design novelty, especially as information technology brings further access to supply chains and schedule logistics.

DESIGN TO FABRICATION INFORMATION TRANSFER

Once a design team has a basic schematic design concept for a building, a modular fabricator can be engaged. If the architectural drawings have not taken into consideration assembly, there will be an exercise in which the modular team develops drawings based on the architectural intent that will allow for a modular build. Here a series of subdivisions occur within the plan and are routinized based on the dimensional capacity of the modular company. This capacity may be constrained by factory logics – how large a box the fabricator can build, or transit logics – how large a box the fabricator can ship, or both. In some jurisdictions, building codes and permitting may also limit the size of a modular unit. For instance, many of the modular projects in the northeastern United States, with its many tunnels and bridges, originate in central Pennsylvania. In this case, size will be limited to boxes that are approximately 16'-0" wide by 11'-0" tall, which considers the 48" to 54" bed and wheel height of a truck, unless a special permit is sought. Unit lengths are generally constrained by a flat bed and can be up to 48'-0" long. That the constraints applied to the design and construction of infrastructural elements such as bridges or tunnels must inform the design of a modular residence that is far removed from this constraint brings the relationship between all aspects of the built-environment into a compelling juxtaposition.

It should be noted in these instances that achieving a maximum length does not always lead to optimization. For instance, a residential bar typology, which efficiently organizes dwelling units on either side of a hallway, will have a maximum depth of about 70'-0" given the depth of natural light penetration that is possible. In this case, it would be more logical to create a unit of about 30–35 feet deep, with the hallway attached on the "back" of one unit, so they are set without having to build a unit demising wall within one of the modules, which can complicate fire rating.

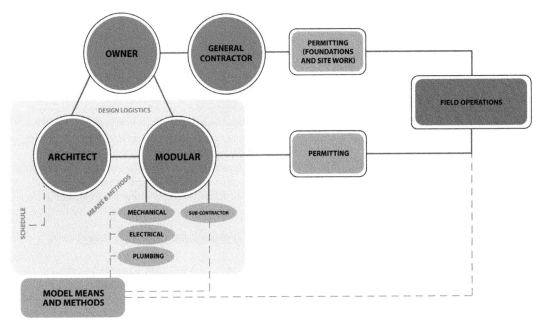

Figure 11.6 GRO Architects, Modular Fabrication with an Information Basis, 2022: By leveraging information modeling, architects can gain scope and impact within a modular building fabrication and assembly schema. Through the 3D design and coordination of intent, the model can capture input from engineering consultants including MEP, as well as construction information and even stacking logics and order. Construction permitting would still be within the scope of the modular company or a local general contractor.

In selecting a modular or prefabricated system for construction, architects expand their scope to consider both operational and delivery logics. On the design side, there is certainly a sympathy between such thinking and information modeling packages, with their libraries of typical wall and floor assemblies being applicable to a standard modular fabrication and assembly schema. Recall there are many similarities between "stick-built" site construction, either light-gauge steel or wood, and modular construction, the most significant difference being the increase in wall and floor thickness to account for the enclosure surface of each unit.

Using information modeling, architects have the ability to link their efforts with those of the modular fabricator, many of them only having two-dimensional capacity in drawing production. Through the model, architects can convey both design intent and some ideas about constructability and manufacture. In these instances, the workflows offer not only an expansion of territory for the architect but also more readily align our scope with that of construction and how our buildings are inserted into sites with various constraints. Simulation routines in the manufacture of modular or panelized systems, and the order in which they are assembled on a site, become within the consideration of the architect. Especially in larger modular assemblages, where higher degrees of design and coordination are required between architect and fabricator, opportunities for the expansion of architectural scope and further definition of architectural *agency* are robust.

PERMITTING AND INSPECTION

Permitting, the approval of a building application for construction by a local municipality or agency, is achieved through inspections at the modular factory conducted by a licensed third-party. In addition to potentially taking scope from the design and site-construction teams, the modular fabricator assumes the responsibility for permitting and inspection. Permitting usually includes the transmission of a printed or electronic set of drawings, which are reviewed by the locality with respect to applicable codes and standards. Construction permits may be obtained by a local general contractor, the modular fabricator, or both.

Within the factory, reports will be generated with respect to framing, electrical wiring, mechanical ducting and venting, and plumbing; as well as fire separation and rating(s) achieved, and sprinklers. Once the modular units are shipped and set on site, the manufacturer will provide these reports to the local municipality for review. In some instances, the actual work may not be reviewed in the field by the municipal inspector.

The lack of vertical continuity in modular systems gives rise to the need for enhanced lateral resistance and figures into the inspection process. Especially in wood modular assemblies, which are stacked by floor or level, shear can be a significant consideration; however, since the modular *parts* are not assembled into *a whole* until they are set at the project site, vertical continuity is usually addressed during stacking to ensure the units will not shift once in place. In many instances, strapping – semi-flexible metal plates – are shipped attached to a modular unit. The straps are fastened to the frame assembly, usually between a floor-to-wall connection in the modular unit and, once stacked into position, are bolted or welded to the unit above or below in the field. The straps, which are similar to those required in high-wind or hurricane prone areas, provide the requisite lateral stability in low-rise structures.

In buildings requiring higher lateral resistance, moment connections are possible by welding or mechanically fastening larger shear elements such as diagonal braces or vertical threaded rods within boxes. These connections are usually steel or aluminum, which affords higher and more consistent tensile strengths than wood.

MODULAR MOXIE

GRO Architects recently worked with a developer on the design of a 79-unit modular micro-housing structure, called The Moxie, built nearby our Nest building in Jersey City. There are several key differences between the two buildings. First, Moxie would utilize more standard "studio" apartments with a base area of 350 square feet, and next, each of these units would have its own full bathroom meeting all the "room" requirements of building and Americans with Disabilities Act (ADA) codes. This building was imagined as a five-story assembly, with a noncombustible base, and four levels of modular construction above.

GRO Architects went through a design and construction exercise with Modular Steel Systems (MSS), a modular fabricator located in central Pennsylvania on the actualization of the units. The desired size of the dwelling units allowed each studio to be achieved in a single module. Unlike Nest which was a panelized light-gauge steel construction solution, Moxie would take advantage of International Building Code (IBC) Type V-A construction which permits four stories of combustible construction, in this case modular wood frame, above a non-combustible base, which was achieved with site-cast concrete with some concrete masonry unit (CMU) block. A three-hour fire separation is required between the noncombustible and highest combustible floor. V-A is a popular construction typology in which the ground floor is given to a parking garage. The urban site zoning for Moxie permitted a five-story structure as-of-right with no parking requirement, therefore the ground floor

Figure 11.7 GRO Architects, Moxie, Jersey City, NJ, 2021: As a modular project, the owner sought to keep construction costs low and utilized PTAC heating and cooling units for the studios. PTACs are required to have air intake directly from the wall behind the unit itself, requiring an unsightly grill for ventilation. GRO sought to minimize the vent's appearance by placing it below a series of projections in the façade, which were used for additional unit space. The projections cast shadows throughout the day, keeping the grills in the shadow path. At the time of photograph, the building was still under construction, but all of the modular units had been stacked in place.

could be utilized for additional dwelling units as well as amenity spaces. This organization set up a "split-level" first floor, where the building lobby and amenity spaces were given ceiling heights of 1.5 stories where additional height was desirable, and the dwelling units were raised 4'-0" to ensure privacy for all windows that were sidewalk facing.

Though the project was always imaged as modular, the development process changed midway through design, following receipt of entitlements. The owner had intended at this time to transfer design development and construction documentation scope primarily to the selected modular fabricator, with GRO taking on a reduced scope to include shop drawing review of modular units. Ultimately the building, which was L-shaped – a less-common building organization for the efficiency of modular and having five of its six facades on or immediately adjacent to lot lines – proved to be too complex for the modular company to detail in-house. The perimeter shape and split-level ground floor also had various egress constraints for a building of this size and density that would have to be met.

GRO was re-engaged for the construction documentation and the information transfer of design data to the modular company. Given the small unit size, the modular fabricator had sought to develop this building as if it were a hotel, only to find regulations for multifamily housing to be more restrictive than those for a hotel, specifically with respect to ADA requirements as all units were required to be accessible.

Information transfer occurred primarily through Autodesk Revit, with units being designed with 2 × 6 dimensional lumber for all façade walls, providing a two-hour fire separation. Unit demising walls were required to have similar fire resistance but, given the modular logic of each unit having its own enclosure, the demising configuration was achieved by two 2 × 4 walls with insulation and a 1" air gap between, taking advantage of the double wall thickness of the modular system. A series of meetings were held with the modular fabricator, and an export of 3D building geometry to 2D vector information was provided for shop drawing development. Design intent was reviewed through shop drawing submittals, which captured all routing for conduit and piping, as well as mechanical ducting. Interior fit-out including a modest amount of built-in components, as well as the kitchens and ADA bathroom, were represented in every unit.

REINFORCED AND PRECAST CONCRETE

Reinforced concrete remains the most widely used construction product in the world. The material has a sort of dual history. It is both a historic material, mixed through the ages, with minimal technical knowledge, of sand, water, aggregate and a binding agent such as Portland Cement as well as a highly technical one. In the postwar years of the mid-twentieth century, a conflux of advances in chemistry and the engineering disciplines allowed the material to be better codified as a prefabricated construction system, aided through the availability of factories abandoned with the passing of the war effort and an immediate need for housing in its path in Europe. It should be noted that the material has its critics, as the binding agents typically used emit carbon during concrete production.

Today precast concrete remains a simple material in its composition and application, while being a highly specific medium – a sophisticated material that can answer a complex array of performance and shape criteria. Innovation has allowed precast concrete producers to advance into both structural precast applications, used in roadway infrastructure as well as building types including warehouses and parking garages, as well as architectural precast applications – in which façade solutions are produced.

ARCHITECTURAL PRECAST
Architectural precast concrete is an outgrowth of the twentieth century and modern means of production; and contemporary developments in materials, manufacturing,

Figure 11.8 Northeast Precast production floor, Vineland, NJ, 2022: A state-of-the-art production facility was recently completed by Northeast Precast at which various types of panel solutions, ranging from roadway infrastructure to insulated precast insulated panels (PIPs) and custom solutions can be produced. The facility is a four-bay production space with gantry cranes and an operable wall on the rear short-side of the building for moving large format panels outside and on to beds for shipping.

and erection procedures have expanded the role of architectural precast concrete in the construction industry.[4] Technology continues to push the complexity of panel design in building solutions, and precast companies are leveraging technology in more interesting ways. As use of building information modeling (BIM) has expanded, producers have developed different approaches to both construction solutions and client needs. Precast providers have become more immersed in three-dimensional production environments; they have moved beyond more general issues of constructability, now favoring this mode of design and production to find ways to optimize all elements and components that go into a precast solution. This might mean lessening required volume of material or optimizing the amount and layout of reinforcing within a panel. In essence, the digital workflow has both expanded and simplified the solution set.

STRUCTURAL PRECAST

Reinforced concrete precast systems have become favorable as modular structural solutions and have gradually transitioned to architectural precast, and façade solutions. The ensuing complexity has required companies like Northeast Precast and Clark Pacific, on the East and West Coasts, respectively, to reimagine pre- and post-tensioning schemas for their panels. For this they have gone to specific software, such as Tekla and Revit, BIM platforms that have plug-in functionality for precast systems. Optimization through finite element analysis (FEA) within the Tekla environment allows precast producers to understand and ultimately reduce material costs, including reinforcing steel and panel volume for full precast building solutions used in data centers and parking garages.

STANDARDIZATION IN THE PRECAST INDUSTRY

As precast concrete production developed from twentieth-century mass-standardization factory models, companies like Clark Pacific have developed a series of "standard" products and production methods. Doug Bevier, a production engineer at Clark, suggests the company explores custom approaches and custom processes for panel design and manufacturing as pursuits require. For thin-shell applications, especially those that are more standard (less "face articulated"), they have adopted a Revit environment. When panel shapes slope in multiple directions, they have explored pretensioned panel designs to meet project specifications. More complex geometric solutions facilitated a move to Tekla, which provides a more integrated modeling environment to explore design and manufacturing options.

Clark Pacific comes to project contracts in multiple ways, and many times this depends on whether the project owner is a developer or an owner-operator, as well as how early precast is selected within a project timeline. The level and time of engagement is related typically to panel geometric complexity or performance requirements with respect to a façade. To this end, Clark Pacific has developed a collaborative delivery approach where their preference is to be very active throughout what they refer to as a design-assist phase, a project and information delivery phase done in collaboration with the design and general contracting teams with the intent of reducing cost and time of construction. The goal is to make things more constructable as well as ensuring envelope performance with respect to fire safety and heat transmissibility. During this phase, project aesthetics are also discussed through the development of a matrix of aggregates for coloration, and other material finishes are also reviewed.

Most of the technical engagement the team experiences during design-assist is with the architectural design team; however, Clark Pacific understands the potential value they bring at a higher level, in which efficiency and cost-savings are brought to the project owner during building delivery and postoccupancy. Solutions satisfy the aesthetic requirements of the design team, and technical details are coordinated with the architects, allowing Clark to remain committed to the "total value of the system." This concept is, of course, important to the project owner, and includes efficiencies in the manufacturing process while achieving the desired effect, as well as panel performance over time.

Since prefabrication relocates a labor force from a job site to a factory floor, performance criteria such as thermal transmission, sound transmission class (STC) ratings and weight performance, must be considered through simulation. Through this process, Clark can estimate energy costs that will be used or saved on an annual basis. The durability and lifespan of a panel or system is also studied with a goal of 60 to 100 years.

PRECAST AND CARBON EMISSIONS

Given all the positive performance criteria and geometric possibilities with concrete, and especially precast concrete, the material's relationship to carbon remains a vexing issue within the larger industry. Many precast producers are conducting internal research on more sustainable, or low-carbon concrete. At the time of writing, carbon reduction, which is achieved by use of a chemical binding agent other than cement, is not driven by building codes. It is understood, however, that reduction is the goal, and something passionately deliberated by the International Code Council (ICC) and American Concrete Institute (ACI). ACI Section 318-19 sets forth building code requirements for structural concrete.

Operational carbon, which is the total amount of carbon emissions during the operational, or occupied, life of a building is commonly referred to in such discussions. It should be noted, however, it is estimated that up to 80 percent of embodied carbon is contained within the structural system of a building.[5] From a performance standpoint, precast concrete does well, but the allocation of embodied carbon, which occurs at the time of manufacture or production, remains proportionately high for precast relative to other building materials. The impact of embodied carbon is measured with immediacy, while the measure of operational carbon is incremental and may be over a 50- or 60-year life of a building. Embodied carbon has been a focus of carbon reduction within architecture, as well as material producers.

The cement content in concrete is what affects its contribution to embodied carbon, so a common approach has been to reduce cement, the binding agent that gives concrete much of its workable strength, by adding other materials. In a typical thin shell application, the volume of concrete makes up about half of the dimensional materials but contributes to about 80 percent of the embodied carbon of the whole assembly, even though its dimensional thickness might be less than 3". Within the makeup of the concrete itself, more than 95 percent of its embodied carbon is the cement content. Several material replacements for cement are currently being tested by the industry:

- Ground granulated blast furnace slag, or captured molten iron slag – a byproduct of steel production
- Natural pozzolans such as metakaolin, a dehydroxylated form of the clay mineral kaolinite; or calcinated clays or shales.
- Silica fume, a byproduct of silicon metal or ferrosilicon production.
- Fly ash, a residue of coal production which is already used in Portland cement to improve performance.

Geopolymer concrete, largely seen as a potential replacement for traditional Portland-cement based concrete, is typically defined as concrete that is created by adding fly ash or slag to the sand and aggregate mix.

As the industry continues to explore cement alternatives, building code will continue to cover design and durability requirements. As alternative mixes come into wider use, it is assumed that ASTM International – formerly the American Society for Testing and Materials – which sets standards governing specifications for cementitious materials will update their guidelines to favor more sustainable mixes. For companies such as Northeast Precast and Clark Pacific, such legislation mandating specific embodied carbon rates is a forgone conclusion, and they are actively pursuing the use of binding agents that reduce the embodied carbon percentage in their products. In both cases, their research and development teams are proactive, arriving at new solutions before they are required.

PRECAST DESIGN-TO-PRODUCTION SCHEMAS

Shape grammar and formwork constraints are regularly derived from a three-dimensional model. A common complaint among precast companies spoken to for this writing is though architects are engaged in façade design, they "aren't necessarily thinking about" the manufacturing and assembly process. Understanding that formwork assembly and reusability, and the order and direction of panel stripping, can provide some operational efficiency, more productive collaborations between precast providers and design teams are emerging. Models furnished by the design team are used for reference, but most times do not include information that is specifically needed for the panel design and manufacture. If the design team is still developing their contract documents the precast provider has an opportunity to share their progress with the architect, which can be incorporated into a "global" model.

Increasingly automated production schemas allow architects to engage precast producers, and subcontractors and fabricators generally, in reciprocally productive ways. Processes leading to novel precast solutions are increasingly digital and bespoke, utilizing CNC capabilities, including multi-axis routers, wire-based foam cutters, and plasma cutters. As automated production means increasingly pervade the construction industry, we find ourselves back at the question Sherry Turkle posed to her colleagues at MIT, which questions how much architects should bother with the technical nuances of how our design-intent is made. As automated production techniques become extensions of our creative output, extensions of our very agency, such questions

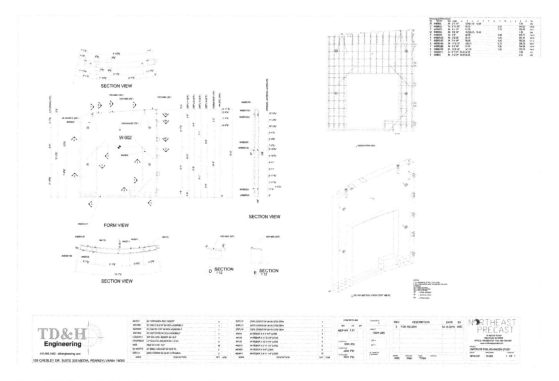

Figure 11.9 Steven Holl Architects, Rubenstein Commons, Princeton University, 2020: Shop drawings for the Rubenstein Commons at Princeton University were utilized by Northeast Precast in the fabrication of the building's curving shell. The building's overall form was generated by intersecting sets of nonplanar curves across the site. The architect was awarded the commission via competition in 2016, and the building was completed in 2022.

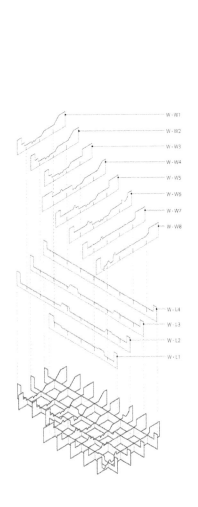
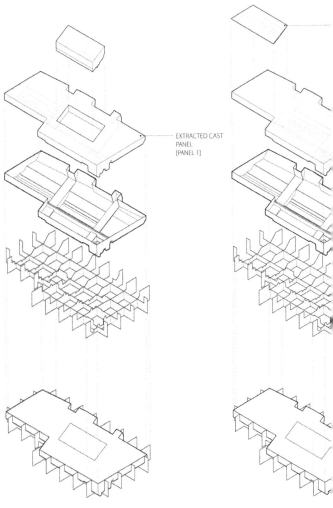

Figure 11.10 Matter, Making, and Testing; Stuart Weitzman School of Design, University of Pennsylvania, 2021: A research seminar titled Matter, Making, and Testing: Designing with Next Generation Precast Concrete is offered to students enrolled in the Weitzman School of Design's graduate architecture program. The course is a collaboration between the school and Northeast Precast, and studies novel panel forming procedures, formwork reusability, and concrete mixes. Student teams are introduced to the material before embarking on a design exercise with Northeast Precast's project managers to produce a series of full-scale architectural precast panels. Student work by Riley Engelberger, Lisa Knust, Lauren Hanson, and Maddy Tousaw.

- FOAM BLOCK OUT
- EXTRACTED CAST PANEL [PANEL 2]
- STEEL FORMWORK
- FORMWORK WAFFLE FRAME
- COMPILED ASSEMBLY

Figure 11.11 Matter, Making, and Testing; Stuart Weitzman School of Design, University of Pennsylvania, 2019: In addition to developing a collaborative digital workflow in which three-dimensional geometry is rationalized into a series of shop drawings, students spend two days at Northeast Precast's production plant, where they assemble CNC-produced formwork, lay rebar, and cast panels.

seem to be playing out in favor of the technologies we increasingly use. Simulations will not only *validate* our work from an efficiency basis, but allow others to see our architectural impact through the aesthetic and spatial implications of carbon-free material and fabrication solutions.

NOTES

1 "The Grandfather of Prefab: An Interview with Carl Koch," *Progressive Architecture*, vol. 75, no. 2, Feb. 1994, p. 62.
2 Rebecca Mead, "Transforming Trees into Skyscrapers," *The New Yorker*, April 18, 2022; https://www.newyorker.com/magazine/2022/04/25/transforming-trees-into-skyscrapers, accessed 29 April 2022.
3 Alana Semuels, "Wildfires Are Getting Worse, So Why Is the U.S. Still Building Homes with Wood?" *Time*, June 2, 2021; https://time.com/6046368/wood-steel-houses-fires/ accessed 29 April 2022.
4 Precast/Prestressed Concrete Institute, *Architectural Precast Concrete*, 3rd edition, First Printing, 2007.
5 Paula Melton, "The Urgency of Carbon and What You Can Do about It," BuildingGreen, https://www.buildinggreen.com/feature/urgency-embodied-carbon-and-what-you-can-do-about-it, accessed November 20, 2021.

IMAGES

pp 178–182, 184, 187, 194–195 © GRO Architects; pp 189 © Northeast Precast; pp 193 © Northeast Precast: Precaster of Record (supplier and installer) Steven Holl Architects, Guy Nordenson and Associates; pp 196 © Northeast Precast

12 | ON PRACTICE

*Nothing is more crucial to a society in crisis
than the ways it chooses to represent itself.*
— Annie L Cot
Neoconservative Economics, Utopia, and Crisis[1] (1986)

How architects choose to use the new efficiencies we are afforded through technology, especially against a backdrop of interconnected phenomena spanning the environment, capital, trends, and our own objective goals, is a complex question. It has been suggested these potentials have driven architectural practice into a sort of *crisis*, perhaps of scope, perhaps of identity. As technologies allow us to readily quantify values ascribed to the former, perhaps they can equally allow us to qualify the latter, speaking to increasing calls to incorporate worth to the architect's creative process.

ON AUTOMATION

A search for the term *automation* in *BIM Design*, which was written largely in 2010 and 2011, yields two instances, and both within a section on Morphosis. Automation today is a big term, through the realization that just about anything can be automated, but also because the impact of automation on labor is now being more fully realized – lauded by some and condemned by others. In design terms, architects have come to understand automation as an outgrowth of digital fabrication schemas that allow us to more easily and affordably achieve actualized geometric (formal) solutions through an interface with computer-numerically-controlled (CNC) hardware. Initially the subject of smaller-scale constructions, the broad use of CNC interfaces has allowed architects to explore formal novelty within increasingly larger-scale assemblies and have enabled further research into human–computer interfaces (HCI).

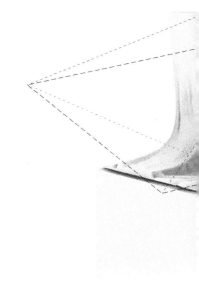

Harvard researcher Andrew Witt suggests that there have been five development periods of computation in architectural practice,[2] all compressed within about a 60-year period dating back to the 1960s and casually related to the use of three-dimensional modeling tools by the burgeoning aerospace industry. Witt concurs that the widespread adoption of computers led to the gradual transition from manual drafting to two-dimensional CAD packages by architects through the 1980s and into the

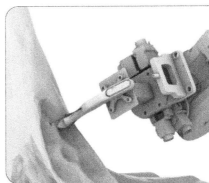

Figure 12.1 Stuart Weitzman School of Design, University of Pennsylvania, 2022: Penn recently launched a Master of Science in Design, Robotics and Autonomous Systems (MSD-RAS) post-professional degree program. The course of study, aligned with the architecture department, seeks to develop novel approaches to the "design, manufacture, use and lifecycle of architecture through the creative engagement with robotics, materials systems, and design computation."[3]

1990s. By then, both solid-modeling packages such as 3D Studio and Softimage, and NURBs modelers such as Alias borrowed from the automotive industry began to take hold. This period allowed for cost-effective architecture-specific modeling packages such as Form-Z and Rhinoceros, which has a user interface and command structure similar AutoCAD, to be adopted by architects. Witt writes that this adoption started in the schools, and with "avant garde practitioners." A deeper integration of these tools with construction brought building information modeling (BIM) to practice in the 2000s, popularized by tools such as Revit and VectorWorks, as well as CATIA, which remained a primary tool in aerospace, but was brought to architecture specific uses with Digital Project, developed by Gehry Technologies.

In the 2010s, scripting tools and graphic user interfaces (GUIs) such as Grasshopper for Rhinoceros and Dynamo for Revit became commonplace. These were primarily utilized in the generation of complex geometry too cumbersome to model for most, but also to automate tasks, making the design process more *efficient*. It should be noted that automation in this regard has existed for some time, dating back to the mid-1980s with AutoLISP functionality being brought to AutoCAD. AutoLISP, or "LISt Programming" allows users with little to no programming experience to automate tasks – stringing together commands to process drawings or geometry. This simple programming language gave way to ObjectARX, based on the C++ programming language.

The progression from modeling to scripting has led to more than simply geometric manipulation in architectural design. Grasshopper, the algorithmic modeling extension for Rhino, was first introduced under the moniker "Explicit History" in 2007, allowing users to more easily access and iterate modeling steps performed on three-dimensional objects. The Grasshopper interface, like Dynamo's, is a wholly other window, in which users drag, drop, and connect Rhinoceros commands for the processing of geometric models. Its interface is intuitive and allows users with minimal scripting experience, but a knowledge of the Rhinoceros command system, to create generative models.

Witt concludes that in the 2020s, architects are increasingly looking into machine learning and artificial intelligence as our "next frontier," but we didn't simply arrive here. The expansion of computational protocols in design has mated nicely with the fact, to quote Neil Leach again, that architects have always "made objects"[4] – that fabrication, in its manual forms, and work with materials have always been the purview of architects and their practice. Such work and research have allowed architects to expand our agency and further interface with those who consider and ultimately construct our design intent.

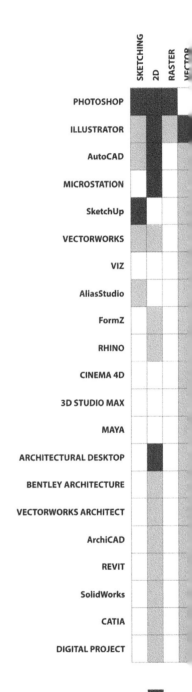

Figure 12.2 Marshall Brown, University of Cincinnati School of Architecture and Interior Design, Forgetting Drawing: The Visualization of Architecture in Digital Media, 2007: Architect Marshall Brown and graduate students from the University of Cincinnati School of Architecture and Interior Design devised in 2007 what they called an "Affordance Matrix" to codify digital applications then-available for architectural design. The matrix allowed for a direct comparison of each program's capabilities, or affordances. The capabilities of a digital medium, along with its limitations, must be considered when matching a software application to design intent.

Automation also emerges from the twentieth century efficiency paradigm of mass-standardization, itself an outgrowth of the Fordist principles that brought the assembly line early in that century. The transition from a mass-standardization basis to one of mass-customization in the twenty-first century was precipitated by technology, and in particular its accessibility, and a *both-and* desire for novelty and efficiency. In this sense, automation has moved from being *fixed*. As manufacturing processes became more automated, they began to align with the rapid prototyping activities of architects at the turn of the century. This partly has to do with an understanding of geometry within the three-dimensional environment of the computer and the code many CNC machines use to form material.

Sharing similarities with low-level programming languages, machine code at its essence drives hardware by providing a series of sequenced cartesian locations – for removing or depositing material – while also setting machine constraints such as the rotational speed of a tool, and the feed rate it moves across a bed. Machine code can be automatically generated by third-party plugins to popular three-dimensional modeling programs and then modified by the designer in code-editing software. While this confluence may have been unintended, it was certainly consequential to architects, already adopting basic coding to automate tasks and becoming more comfortable navigating immersive three-dimensional digital environments.

Our future lies with some level of human–robot cooperation, through automation more specific to our human condition today. The architect's history is filled with the ethical and social implications of our work, especially in terms of how we engage those who build and occupy our work, but operate outside of the traditional scope of design. There remain levels of disparity in the structure of how our buildings are constructed. While construction processes are not neutral, it is worth speculating about how things are made, the processes that are undertaken, and how those involved in those processes are engaged and compensated for their work. Means and methods have traditionally been outside of the architect's work and scope, they are in most contracts the purview of the general contractor, but if machines and other intelligence are contributing to the construction of our design intent, how do we engage these concepts?

ARCHITECTS AND THE DEVELOPMENT SPACE

Real estate development, once solely the purview of policy makers, planners, and attorneys, has become increasingly viable design territory for architects. Now understood as part of a "value adding" equation to such endeavors, the design intent of architects, and procedures and processes we use to bear them, have found increasing use across various stages of development. These include visioning and creative thinking at the outset of any development, the geometric response to various bulk and zoning or land-use criteria during a building's schematic design phase, the application of those metrics to various funding, building, and costing models during construction documentation; and the utilization of construction phase models to synthesize data from earlier development stages into an actual building. The codification of such a continuous workflow, aided by new technologies, ensures development goals are met while allowing for a common language across disciplines and multiple stakeholders, bringing further utility to constraints and data that drive what can be seemingly discreet endeavors.

The relationship between architects and developers in the late twentieth century was a curious one. The trend had been, and still is to some extent, that development projects were maximally *efficient*, that is, as lean as possible in terms of built space and the cost of achieving it, while maximizing portions thereof that could be monetized. Numerous architects scoffed at this relationship and the design process it bore, which was seen as severely limiting that process and the architect's vision in general. Many of these experiences ended badly, with the architect and developer, who *is* the client in this relationship, having differences in opinion about what matters, where money should be spent, and even project aesthetics. The common thinking on "our" side was that the developer was an over-grounded profit-seeker, who would many times overlook more qualitative opportunities within projects if they could not be quantified economically. The reverse thinking was the architect was an "artist" who has an unrealistic relationship with conditions on the ground, which drive development logic and the entitlement process.

Peter Hendee Brown points out that a romanticized model of architectural practice – if it ever existed at all – is far from normative, noting that more architects are working in fewer, larger, firms and agencies while the work of the architect is becoming increasingly commoditized. "These trends have highlighted the conflict between the romantic image of the architect-as-artist and the reality of the architect as a provider of professional services for fee-paying clients who require effective and economical technical and aesthetic solutions to their everyday problems."[5]

These interactions, however, are no different traditionally, and no less fraught, than those between architect and general contractor in the sense that both parties want a positive outcome, however, articulate their goals differently, linguistically and perhaps materially.

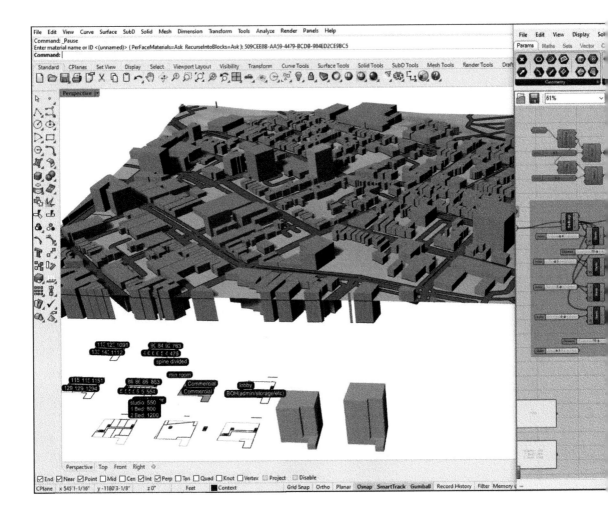

Just as building information modeling and related preconstruction technologies can be used to bring together the assorted parties charged with realizing a building, so can they couple design teams with developer clients who have had issues with intent and articulation. The goals of a developer client involve reaching certain optimizations when speculating on a parcel. This includes, broadly, obtaining control of the asset for the lowest cost, while maximizing the amount of monetized space that can be constructed on the parcel. This might mean achieving the highest residential density, or commercial square footage, while reducing areas given to circulation and services – the so-called "loss factor" of spaces that cannot be otherwise monetized. The mechanics of such capital-forward motives have disinclined many firms from participating in the "developer space" and have frustrated others that attempted to work within such economic constraints. Still, these limits are exactly that, *constraints*, that in a design sense are not so different than programmatic requirements or construction logistics, these are limits within the project scope that architects must engage during design development.

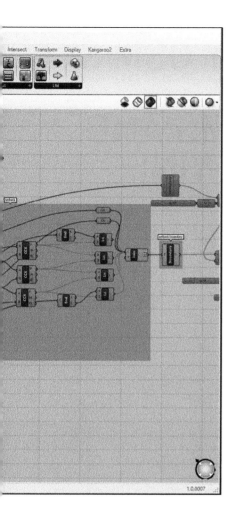

Figure 12.3 Real Estate Development for Architects, Department of Architecture, Weitzman School of Design, University of Pennsylvania, 2021: Architects are increasingly utilizing generative design tools and applying them to traditional scopes such as zoning analysis. With a GUI such as Grasshopper, building constraints including site perimeter, yard setbacks, and building heights can be implemented to generate massing solutions that can be further queried. By exporting data to a spreadsheet, the program becomes the basis of a pro forma where unit costs and other variables can be attached to geometric attributes such as area and loss factor to generate total cost. Such is an example of how architects can add value, an aspect important to clients, while increasing both scope and agency. Such work has also been explored by students in the author's Introduction to Real Estate Development for Architects seminar, taught at the University of Pennsylvania.[6]

While the majority of thinking about BIM has been applied to the relationship of architects, and their models, to construction activities, the model as a digital parametric assemblage lends itself to pre- and schematic design phases as well. While each subsequent phase produces a more detailed design proposal, it's important to recognize how this information is used, and parsed by stakeholders involved in a building realization process. The utility of a model in clash detection, spatial resolution of components, and three-dimensional visualization to a contracting team has been proven, as has data sharing, either as three-dimensional geometry, two-dimensional drawings, or numeric information. It is the latter that also finds utility with developers in pre- and schematic design phases, where the generative capacities of a BIM can be used to receive inputs with respect to zoning information including height, coverage, and setbacks; and drive outputs in terms of quantities and takeoffs that can be used with a pro forma or other financial and risk calculations.

NOTES

1 Annie Cot. "Neoconservative Economics, Utopia and Crisis." In: Zone. Vol. 1/2. New York: 1986.
2 Andrew Witt, *Formulations: Architecture, Mathematics, and Culture* (Writing Architecture) (Cambridge: MIT Press, 2022).
3 https://ras.design.upenn.edu/ accessed 10 May 2022.
4 Neil Leach, "Digital Tool Thinking: Object Oriented Ontology versus New Materialism," *Posthuman Frontiers* (2016).
5 Peter Hendee Brown, *How Real Estate Developers Think: Design, Profits, and Community* (University of Pennsylvania Press, 2015), p. 95.
6 Students who specifically analyzed the illustrated site were Peik Shelton and Joo Young Ham.

IMAGES

pp 198–199 © Weitzman School of Design, University of Pennsylvania; pp 200–201 © Marshall Brown Projects Credit: Marshall Brown. Student participants: Luke Field, Preeta John, Adam Koogler, Thomas Rudary, Jacqueline Squires, Eric Stear, and William Yokel; pp 204–205 © GRO Architects

13 | ASSEMBLY OSM'S MODULAR PLATFORMS AND DIGITAL TWINS

Assembly OSM's stated goal is to deliver high-rise buildings in dense urban environments through affordable offsite manufacturing schemas. In doing so, it has created a digital platform to reimagine the entire construction process, utilizing digital design and fabrication processes to respond to a variety of conditions in the city.

Assembly's platform leverages CATIA's 3D environment to challenge a construction industry that has not made significant advances since the Second World War, where manufacturing was pointed to producing housing, especially in Europe. Assembly grew out of the design research being conducted by SHoP Architects, where a focus has been on how construction can be streamlined through a digital-twin platform. The platform manages everything from envelope and zoning, to specific building design, to financing models, to ultimately tracking the progress of the design through a financial market analysis (FMA) and on-site assembly. This work started with SHoP's Dean Street project in Brooklyn, where the firm embarked on an ambitious high-rise modular construction schema that was met with significant hurdles during construction. The firm acknowledges this was somewhat of a blemish on its advancement of a design to manufacturing workflow.

Assembly employs over 30 designers and technicians with experience in direct-to-manufacturing schemas. Advisors include SpaceX and Tesla. A retired CTO of Boeing is an investment advisor, and the company has raised over $20 million in seed funding. It focuses on addressing housing in our cities as land acquisition and construction costs have risen to the point of both pricing out consumers and challenging the feasibility of development projects. Assembly seeks to leverage the robust design research by SHoP to bring a highly designed and efficient product into a territory that has traditionally been relegated to low-bidders and buildings that have both quality-control and cost challenges. While many architects have taken on the challenge of well-designed and cost-efficient housing,

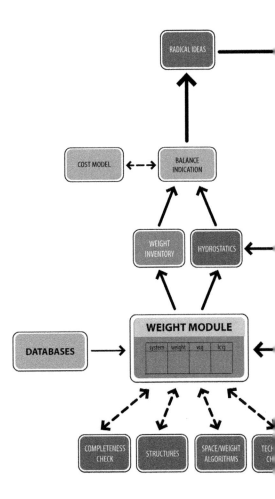

Figure 13.1 GRO Architects, Design Building Block, 2022: The "Design Building Block" method as developed by D.J. Andrews and others, and adopted by the shipbuilding industry, posits that discreet parts are developed at the onset of a digital design process and iterated through simulation and analysis both discreetly, and as an assembly. The workflow establishes a series of platforms, including "geometric definition" in which hull forms are optimized, and the "weight module" in which mobility and stability are studied. In many cases, the design and fabrication of watercraft occur at multiple sites concurrently, allowing design team members to collaborate in an immersive digital environment. Such a decentralized model of design and assembly has been utilized in the aerospace industry as well and has implications for the offsite design and assembly of buildings and objects designed by architects.

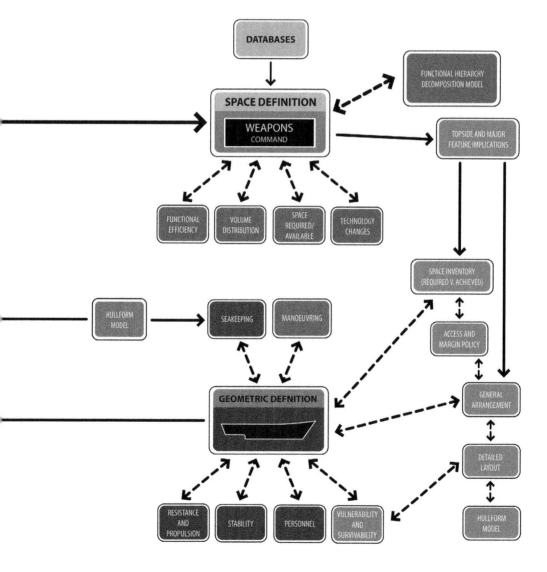

many efforts have been marred by the effective delivery of design-intent, and prefabrication schemas such as modular construction have not had as successful a track record as hoped.

Assembly seeks to differentiate itself from competitors by taking on the challenges of assembling prefabricated parts into a (whole) system that is craned into place. This is analogous to the CATIA technology they leverage as individual parts are assembled into products within that digital environment. The level of calibration, tolerance, and ultimately code-compliance demands that Assembly rigorously controls in their assembly process occurs through this technology.

To achieve these goals, it has sought to create a direct supply chain with established manufacturers to include all raw materials as well as labor. This expanded team provides materials and expertise for a series of building modules, including a structural steel chassis, façade, floor "cassettes," and bathroom and kitchen (wet) pods. To do this effectively, Assembly has devised a tier system in which these suppliers exist within Tier 2 of the process, with Tier 1 being onsite construction. Cimolai[1] is one of three chassis suppliers used to manufacture prototypes currently. Cimolai operates worldwide in the design, supply and erection of complex steel structures ranging from bridges and stadiums to architecturally complex buildings. Their experience leads them to understand the tolerances that need to be maintained.

The steel chassis of the Assembly module is flat-packed, and tolerance requirements are met through an adjustable fastening mechanism that allows for levels of alignment, permitting components to be fastened for shipping before being finally locked into place at a Tier 1 site. The firm has completed a series of stacking prototypes with a rail system that locks components together.

A key to the Assembly model is that all components for their system are produced, manufactured, or assembled by companies that already exist. Novelty is created through the digital management across this supply chain. Such a model is already being utilized by the automotive and aerospace industries where prefabricated components come together on a project. In this model, Assembly retains ownership of intellectual property.

Manufacturers from SpaceX and Tesla collaborate with architects from SHoP or those who have trained at SHoP. Other team members include mechanical and structural engineers, as well as industrial designers from companies such as Frog Design who are providing interior finish concepts. Arup is working with the Assembly team on acoustics, mechanical systems, and fire protection of the modules.

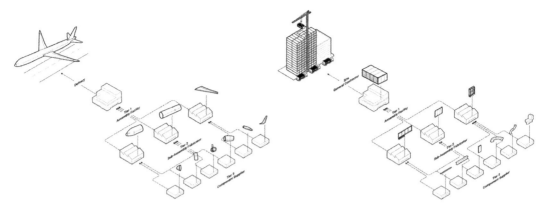

AEROSPACE INDUSTRY   ASSEMBLY OSM

Figure 13.2 Assembly OSM, Module Production, 2022: The Assembly process involves a series of three tiers, for which various contractors and suppliers can be prequalified. The process starts with a Tier 3 Component Supplier for processed materials, a Tier 2 Sub-Assembly Fabricator for initial fabrication, and Tier 1 factory for assembly of the overall module within a local assembly space. The completed module is then transported from the warehouse to the project site for erection by crane. Such a platform model is already utilized in the shipbuilding and aerospace industries.

Assembly customized the CATIA environment through scripting to facilitate their stated goals. All building components are designed, and exist, virtually for manufacturing, so there is no representation necessary in the platform. The "key IP" is in the locking mechanism for the structural steel chassis and the façade attachment. The system is designed to accept a wide variety of façade materials and organizations, allowing the Assembly system to function somewhat as a backdrop for an architect's expression. The novelty of the design is in the way it is made, and affords both cost-savings and speed of erection. While there are limits to the floor plate configuration due to the modularity of the system, variability and a level of customization can be achieved in façade design. This is a crucial aspect of the assembly system – that a façade designed by others can be accepted so long as the fastening system to the module can be accommodated, allowing for the Assembly system to exist "behind" a wide variety of buildings. The utilization, and intelligence, of this system challenges the traditionally bespoke nature of the architectural object, which historically has been seen as a one-off that takes years of planning and preparation prior to execution.

Tier 2 fabricators manufacture individual building parts or prefabricated assemblies of components which are shipped to a Tier 1 builder, a general contractor who assembles the final system on site. Christopher Sharples, a partner at SHoP and the co-creator of Assembly with his brother William, thinks a key to the system is that none of the products are *sole-sourced*. A developer cannot simply hire a company to provide all of the parts or materials to build a high-rise, and the Assembly system allows multiple manufacturers or suppliers, all of which have been prequalified to communicate with the digital platform, to bid on these components so a competitive cost can be achieved. Such competitive bidding is also a way to navigate the complexities of material costs in robust markets.

In such a distributed model of production, shipping and transit costs can be calculated along the supply chain. For instance, while steel erection costs of a Tier 2 supplier in Europe may be more cost-effective than one in the Americas, the freight costs may prove prohibitive, which would be reflected in Assembly's platform.

TIER 1 SYSTEM

Assembly has determined that its system is far more cost effective than a traditional modular construction mobilization, which requires time and resources to construct a factory to mass-produce components. Sharples feels Assembly can spend what amounts to a fraction of the cost to create such a factory through the investment of local high-bay warehouse spaces to further assemble various components that are supplied by Tier 2 companies. The proximity of these spaces is specific to key regions that Assembly serves.

Assembly is currently working with two North American fabricators to further streamline this process. The level of accuracy demanded in its system has been a challenge on the factory floor, but Assembly has provided specifications through scanning technology that allows manufacturers and fabricators to become more comfortable managing tolerance risk.

For heating and cooling, a mechanical module, or "pack," is attached to the requisite ADA bathroom, which is typically planned deep in a dwelling unit, where natural light penetration is not required. This positioning allows the mechanical module to be serviced from the hallway, as access is a key component to maintenance in apartment buildings. Any ducting already present in the unit module is connected to the air handling unit in the mechanical pack at the time of installation, occurring at a Tier 1 location. Module sizes can vary, but a standard width is 13'-6" and height is 11'-0" tall, the latter being for the maximum under which special permit transport is required, and travel over bridges or tunnels is a critical consideration in the metro New York housing market, as discussed in chapter 11. Module widths can be as small at 12'-0" wide and 10'-6" tall or large as 14'-6" wide and 11'-6" tall. Module lengths are between 24'-0" and a max of 40'-0".

Assembly has used this research and development period to understand regulations with respect to zoning and building code, basing their work on New York City Department of Building (DOB) standards. Assembly acknowledges that many project funding sources are tied to the construction permit process, so understanding how their system performs with respect to DOB standards has been a critical benchmark in their work. Life safety and fire rating standards are also significant considerations to the modular process. There is no gypsum board on the interior of the Assembly system; instead a prefinished

Figure 13.3 Assembly OSM, Module Production, 2022: The Assembly team relies heavily on the CATIA platform for the design and virtual assembly of its module. The firm operates the module as a digital twin within the three-dimensional environment.

panelized system using MDF provides both fire protection and the finish interior surface. Joints are expressed as reveals inside the dwelling units.

Façade systems have always been a vital component to SHoP's work and have been central to the design research undertaken by the firm. Starting with one-to-one scale exercises such as A-Wall, designed and fabricated as a traveling exhibition system to *Architect* magazine in 2000[2], the firm has sustained an interest in both novelty and efficiency in imagining wall and exterior cladding systems. The firm has been able to scale such efforts to projects including the Barclay's Center in Brooklyn, completed in 2012.[3] These efforts took an international turn, with the manufacturing, in Cape Town, South Africa, of façade components for the Botswana Hub.[4] Digital data produced via the CATIA platform in New York was transmitted to fabricators in Cape Town, where they were able to form building components, which were in turn shipped to the site for assembly. This workflow, and the corresponding knowledge transfer to a worker population, proved to be successful and forms the basis for Assembly's larger project.

William Sharples feels SHoP has proven these concepts through the digital transfer of design intent and with Assembly is focused on creating a niche in the US urban housing market. Sharples feels the progression of this work is logical in that it marries an assembly process to a schedule that is put forth by the architect.

This building intelligence also allows the company to "test fit" Tier 1 sites, understanding how the floorplate and configuration of existing warehouse spaces can cater to their needs. Again, an important aspect of the company is that they do not need to create specialized

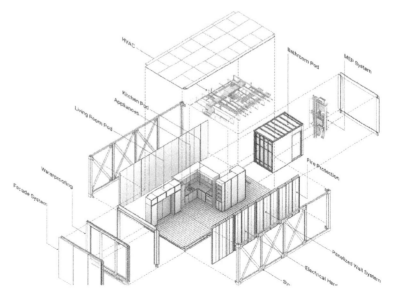

Figure 13.4 Assembly OSM, Module Production, 2022: The Assembly platform captures not only the material fabrication of building components and modular units, but also the procurement of those materials and the delivery of those modules. Assembly goes beyond a typical scope of architectural work to include vendor identification and selection, and process and logistics tracking.

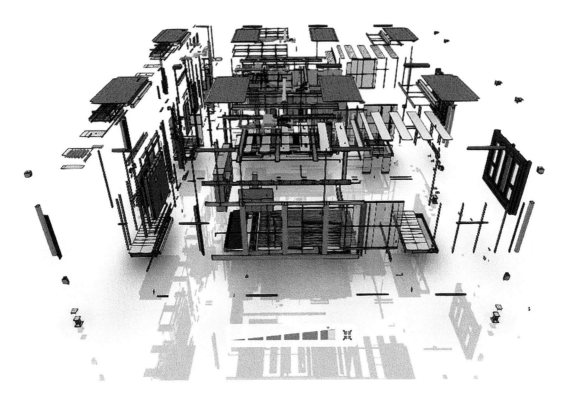

facilities for module production. The constraints imparted in the digital system allow the firm to understand how an existing facility with proximity to any construction site can be adopted as a Tier 1 location. The firm uses virtual reality to this end, understanding through simulation how their manufacturing system can be organized to maximize production of modules, with a goal of six completed modules per day. They refer to clusters of equipment and workers within these Tier 1 facilities as "marriage stations", where components from Tier 2 suppliers are ultimately mated and prepared for transport to project sites.

It's critical to point out that the entire system is tested virtually, simulated in the model, to understand total aspects of the system from the achieving of goal tolerances in joinery to efficiency of module assembly within a Tier 1 location. Another aspect of the Assembly process is a requirement for two separate building contracts prior to investing in a Tier 1 assembly space. In order to receive regulatory approval, the company has gone through the permit process and has achieved New York City Board of Standards and Appeals (NYCBSA) approval for a seven-story building there, including life safety and fire resistance authorization for the structure. The company realizes that fire resistance ratings are the most stringent requirement of the overall system, as opposed to resistance requirements for individual components such as the steel frame. Sharples refers to the fire resistance approval process as a sort of "space race" with competitors and feels the Assembly system is well positioned given the intelligence of their product.

Figure 13.5 Assembly OSM, Module Components, 2022: One of the limitations of traditional modular assembly is the inability to provide adequate shear resistance for lateral loading. This limitation has restricted the height of modular structures to mostly low-rise configurations. With Assembly, SHoP hopes to achieve taller, and denser, buildings more appropriate for urban environments. The ability to stack the Assembly modular system, and its ability to resist horizontal loading, has been rigorously tested.

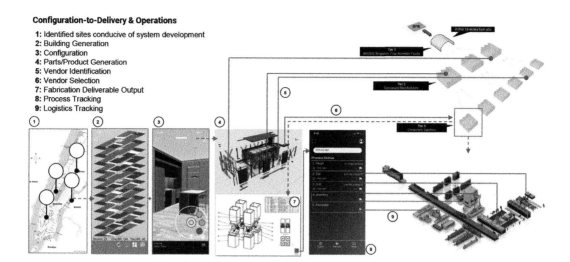

Figure 13.6 SHoP Architects, Configuration-to-Delivery, Harrison, New Jersey, 2022: A significant part of the Assembly logic is the configuration of building products to be delivered to specific site installations. The goal has allowed a franchise-type attitude, where the module can be produced in multiple locations given specific warehouse and vendor constraints are met.

The system has been simulated through structural performance criteria to a height of 30 stories on the East Coast, with the West Coast being about two-thirds of this height due to more stringent seismic requirements. Such simulation is important to understand how the system can be tailored to differences in building codes around the United States. The structure receives moment resistance from a Vierendeel-type system, which is spliced together at the project site. Such innovations mean the Assembly system is appropriate for various dwelling types within the housing and hospitality markets (hotels, short-term rentals) as well as emergency shelters that can be adopted to FEMA standards. In support of the latter, SHoP is already working with the US General Services Administration (GSA) on the design of multiple embassies in different parts of the world.

All of this is made possible through the firm's ongoing interest and development in parametric modeling. The CATIA platform provides the robust environment where multiple sets of constraints can be analyzed given a geometric solution and simulated using virtual or augmented reality. This brings Assembly back to the Digital Twin concept, where each of its material solutions are borne through a virtual counterpart that goes through a rigorous testing and simulation process. The firm no longer relies on representations for others to interpret, which is central to both their system's success as well as the expanded territory within which Assembly hopes to position the role of the architect.

As Assembly expands its ability to license its system to others, a Revit>CATIA workflow has been established for the purposes of analyzing the appropriateness of a given geometric solution to their system. Their platform verifies the solution with the Assembly manufacturing and fabrication system and offers ways in which geometric solutions by others can be mated to it.

In many cases, this has less to do with unit layouts, which can remain the design intent of other architects, than with the organization and spatial relationship of circulation elements. These include stairs and elevator cores, which have both a vertical relationship through a building as well as code-related dimensional criteria within a floor plate. For the Sharples' this is dictated by the manufacturing process. Assembly does not wish to reverse-engineer its system based on building code, but rather understand how code constraints combine with other factors to affect the performance of their system. This idea is consistent with a concept elaborated earlier in this writing, that architects should increasingly focus on raising questions as opposed to setting out to solve discreet problems. Codes evolve, and Assembly seeks the flexibility to take on new intelligence or data, adapting their system.

ASSEMBLY'S BUILDING FUTURES

Architects have always had some relationship with the future, including working on scaled physical models before the codification of tender drawings, making fanciful illustrations of some distant time, or integrating consultants and virtual execution of a complex construction project within a building information model. In all of these, we have speculated on possible outcomes increasingly honed to a more specific actualization through the adoption of digital technologies. As such, it is more revelatory to describe this relationship with the future more specifically as a relationship to time, or temporal activities as they advance in a not-so-linear manner from design concept to realized building. Temporal relationships, whether the coordination of multiple trades within a project site, an understanding of a construction schedule, or managing material change to a project, are again the purview of architects and designers. Lost in the twentieth century to an expansion of the design-to-build team, with the advent of construction managers and owner's representatives, the architect's agency has once again expanded to the realm of building actualization and more specifically the various construction and manufacturing activities that support such actualization.

John Cerone, a principal at SHoP who established the Assembly digital workflow, feels the company's work is not about technology per se, but adopts a speculative position on how technology will impact future practice, especially as it relates to engagement with environmental concerns. Next generation thinking involves large-scale fabrication, though automated or semiautomated means, digital modeling 2.0 of sorts. SHoP's success with project delivery relates to this making. Computation for Cerone is no longer solely used by architects for formal novelty, but for solving

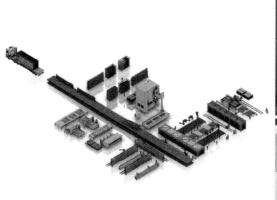

Figure 13.7 Assembly OSM, Process Automation, Harrison, New Jersey, 2022: By leveraging the power of the CATIA environment, Assembly has not only designed a module, but an entire logistics operation behind the internal assembly those modules, and their connection to other modules.

environmental problems through larger-scale simulation and bespoke customization. The organization of design intent per individual preferences shared through a game interface is also easier for stakeholders to understand. Building futures will involve a more comprehensive engagement with building construction while catering specifically to individual, or singular, demand within a total solution set.

Prior to SHoP, Cerone worked for a firm functioning as Architect of Record for the Akron Art Museum, a Coop Himmelblau project in Ohio. He recalls the workflow, which involved receiving a model and generating construction documents from it, a process he refers to as a "source of truth." At the time, 2006, the team's efforts were organized around a Rhino model, where design input took place. Export-type tasks were executed using macros in RhinoScript to facilitate a smooth translation from three to two dimensions. Large-scale physical models were built for design team members and other stakeholders to review.

The museum was a large project, and it was necessary to separate building aesthetics and organizational complexity from a process standpoint. But the design/modeling team was only composed of four or five designers, and the drafting team, which received design data from the modeling team, contained even fewer members to produce the drawing set. Changes needed to be manually tracked, and were time consuming. Though Cerone enjoyed the immediacy or directness of this process, he challenges the efficiency of it in Assembly's work today.

While BIM systems have become more robust, this augmentation has made their operation more complex, bringing what Cerone refers to as a "layer of abstraction" to the process. Assembly is determined to bring directness back to design and construction, both through advances in the modeling protocols they use, as well as through augmented reality (AR) tools that convey to design team members and other stakeholders design intent in real-time virtual environments. The level of control mandated within this virtual material-geometry environment, with part-assembly connections and interfaces within building systems, requires a level of fineness in the model database that can be understood at a series of scales.

The digital twin, which is central to Assembly's approach, is not really a single model or object but a series of data and systems that can be engaged in a variety of ways and at a variety of scales. This brings a level of abstraction to the process that needs to be understood within the workflow, specifically by the design team. Assembly has found ways of bifurcating virtual information based on various tools and interfaces it needs for design development, relying on Autodesk's Revit to "contain" three-dimensional composition and output drawings.

Figure 13.8 Assembly OSM, Prototype Stacking, Harrison, New Jersey, 2022: As a stated goal of the company is to achieve dense high-rise structures within urban environments, lateral load testing within both virtual and actual environments has been undertaken. Assembly believes that their modular product will be competitive in the 8- to 16-story market for housing.

Figure 13.9 SHoP Architects, St. Nick's, New York, New York, 2021: As a proof-of-concept study, Assembly OSM is borrowing a project from sister company SHoP Architects to test the effectiveness of the module system within a mid-rise configuration. St. Nick's is a 100-unit apartment building with an anodized aluminum and glass façade. Assembly worked with SHoP to arrive at the unit mix and interior layouts and completed a feasibility study with a general contractor on a cost comparison for site construction and the Assembly OSM system.

Figure 13.10 Assembly OSM, Chassis Assembly, Harrison, New Jersey, 2022: The core of the modular unit is a moment-frame steel chassis that establishes the size and scale of overall configuration. Assembly is studying several variations of the chassis that yield different unit sizes, all of which are understood within shipping and delivery routines.

Assembly looks for ways to make the geometric modeling tools it utilizes more inclusive in this process, as a way of engaging the immediacy of the design work, linking it back to the directness of a more analog physical modeling process within the digital twin. Cerone acknowledges that the digital twin concept has existed in other industries for some time. Abstraction, however, has been a useful technique in the architectural design process specifically because architectural teams tend to be less diverse disciplinarily. In product design, or the aerospace industry, team members have more varied and specific training and roles, such as soft- and hardware engineers and information technologists.[5]

In Assembly's use of digital twins, they seek to make more intuitive geometric and material decisions while allowing space within the workflow for others traditionally outside the architectural discipline. In Assembly's platform,

architects work on a problem (housing) that has been central to both architectural discourse and practice, and creative novelty occurs both within their product and their approach. Unlike a conventional design process where engineering consultants are engaged following a formal solution put forth almost entirely by the architect, Assembly engages a variety of specialists at the onset to design a base product – the modular chassis itself with all the different connections and systems it needs to function, as well as to customize this solution based on specific sites or product requirements. They are combining aspects of the architectural tradition (e.g., siting), with best practices of established manufacturing and production models. As such, they are relegating aspects of control within the process by relying on the agency of others, while expanding the idea of architectural control over the entire design and construction process.

Another critical aspect of the Assembly model is its engagement of standardization practices manifested in postwar twentieth-century production models with limited use of customization, which ensures Assembly's modular system can be broadly brought to market while allowing for bespoke difference. Most offsite manufacturing technologies engaging the construction industry, including modular and panelized building, rely on standardization to be profitable, a sort of volume-based economic approach. Assembly's product is at once within this spirit yet speaks to the customization possible with twenty-first-century manufacturing methods.

This follows a trend toward a more immersive digitalization to allow design teams to arrive at more specific material solutions. SHoP, and more so Assembly OSM, are invested in these physical-by-virtual operations, especially in the way they are interested in affecting the near-future. Moving to a more holistic model-based environment, both in design stages as well as within the job site, through LIDAR/3D scanning and photogrammetry, offers opportunities to track construction progress within the model and links that progress not only to design documentation and intent but also financial and requisition information as well as schedule. SHoP has already used scanned data to make decisions[6] where technology was utilized at different scales of production, from the local understanding steel tolerance in a panel-cladding system to the simulation of global construction progress.

Assembly is a natural progression of the SHoP model, a sort of DIY approach honed by both experience and mistakes. The firm goes beyond the traditional agency of the architect, citing Boeing as an example of future practice, where the model is central to design decision-making, testing, and ultimately operation. The firm also

Figure 13.11 Assembly OSM, Mechanical Module, Harrison, New Jersey, 2022: Units are heated and cooled via a mechanical module or "pack" located adjacent to the requisite ADA bathroom. The mechanical system is located deep in a dwelling unit where natural light penetration is not required. This positioning also allows the mechanical module to be serviced from the building hallway, as access is a key component to maintenance in apartment buildings.

realizes that there is no "best tool," fostering a curiosity that has brought a broad suite of digital tools, as well as programming and plug-in development into its design workflow.

The early digital design operations that gave rise to SHoP's body of work remain present in Assembly's digital twin in a more automated way. The creation of splines to yield a surface are parametrically controlled and *control* global composition. The currency of these operations lies in a specific local finesse, where, for instance, a series of clips to attach a façade system to secondary steel are constrained so any update to the global composition can be processed automatically. This workflow is bidirectional, as specific material thicknesses or tolerances of those clips will limit the extent of manipulation of an overall surface to prevent performance failure. Controlling geometry through local inputs in early projects led to the firm's interest in modeling fastening components, a more systems-based approach again seen in the Boeing model.

Where Assembly is notable is also in their augmented collaboration with fabricators, and within an organizational idea of their work beyond the digital twin – beyond the model itself – to include the arrangement of teams and hardware, the human-machine interface, of the shop floor where assembly occurs. This reappropriation of work, both digitally and actually, allows the architect to take responsibility for the means and methods of production. Recall such a model has traditionally been avoided by the profession due to the inability to assign error or fault within the messiness of a construction routine. The conservative nature of the industry a contractually a delineated way for blame to be assigned. This has

Figure 13.12 Assembly OSM, Façade Assembly, Harrison, New Jersey, 2022: The façade panel system is designed to maximize daylighting but is customizable so architects utilizing the Assembly module system can design a building façade specific to the site and configuration. Depending on façade complexity, it could be applied in the factory setting or in the field once module stacking is complete.

implications for insurance companies, whose protection of those involved in the design and delivery of a project is based on such delineation, as well as limits to the traditional agency of those within architectural and allied design fields.

To this end, Assembly has been imagined as a start-up company, with its own capital and financial backing – insulating it, for the time being, from this traditional messiness while allowing it to disrupt traditional contractual models. The company is set up to follow the behavioral change that SHoP has embraced. By "evolving the model," the company, in a more automated manner, still creates the traditional two-dimensional drawing set which remains *the* contract deliverable. It draws on the progression from modeling to scripting in the execution of traditional architectural scope while augmenting that scope to include building delivery activities. A path to this augmentation has not existed in the conventional sense, presenting the opportunity to create software and routines that bring elevated control over geometric development, while establishing new protocols for visioning and interaction.

THE INDUSTRIALIZATION OF PROCESS

Cerone sees all of this as an important aspect of the Fourth Industrial Revolution (4IR), the industrialization of process, specifically mitigated within a virtual environment. For Assembly, this means understanding material flows through an immaterial environment, engaging supply chains, and ultimately disrupting them, to ensure materials utilized in their module, specifically

Figure 13.13 Assembly OSM, Module Interior, Harrison, New Jersey, 2022: The design team is very conscious that the Assembly module will have to compete with site-constructed apartment products and have staged a unit in their production facility in Harrison, New Jersey as a proof-of-concept.

structural steel for the module chassis, is located and procured cost effectively and delivered for fabrication. In this sense, Assembly is not a modular company but a process and logistics platform focusing on the material delivery of architectural design concepts. It delivers buildings and has identified a specific need within the mid- to high-rise housing market where a distributed supply chain allows for the procurement and assembly of components that economically challenges the traditional quality, schedule, and cost of a locally site-constructed solution.

The Assembly proof-of-concept "product" currently exists materially as a flat-packed modular building solution, but Cerone allows that other solutions could be simulated, sourced, and procured through the platform. As Assembly's output is directly tied to the flows of materials, data, and capital that exist within the global supply chains the platform digitally interfaces, it has an embedded logic that allows the firm to rethink the material basis for their projects.

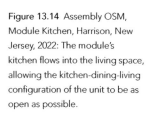

Figure 13.14 Assembly OSM, Module Kitchen, Harrison, New Jersey, 2022: The module's kitchen flows into the living space, allowing the kitchen-dining-living configuration of the unit to be as open as possible.

Figure 13.15 Assembly OSM, Module Interior, Harrison, New Jersey, 2022: Bedrooms and service spaces, including ample storage closets, are accessed from the main living space, consistent with most urban residential products.

This flexibility brings the company back to value, and how value is produced. While architecture has traditionally been considered a service industry – we provide a service by designing structures for our clients – this level of material engagement is not something the modern profession has seen. Assembly's value lies as much in its capacity to disrupt this tradition as in the capability to innovate in terms of building delivery. Cerone sees the long-term relationship between SHoP and Assembly as akin to that of SpaceX and Tesla – two companies that have a completely different product set, but are heavily leveraging shared technology to grow.

Public perception positions Assembly as a technology company where its ability to disrupt is more acceptable than it is within AEC or the architectural space.

FOR REFERENCE ONLY?

Risk and liability are reconsidered within the Assembly platform. The transparency afforded within a model-based environment should inherently reduce risk, but ownership and responsibility standards still have the capacity to silo trades, leaving highly detailed models relegated to a referential status. Since the digital twin contains information about material assembly and operation, it allows for a far more comprehensive understanding of both process and functioning, meaning that *means and methods*, the process of construction relegated to the contracting team and removed specifically from the scope of the design team, becomes part of model delivery. The interpretation required in a traditional building abstraction, such as a set of two-dimensional drawings, is removed, and metadata within this environment also ascribes ownership, literally and figuratively, to specific team members, creating a path for specific questioning and action. A traditional RFI procedure – where a series of questions is transmitted to an architect or general contractor who is required to parse them and share with the appropriate team member – would no longer apply. This ultimately gets back to assigning value.

Authoring tools such as phone apps, have become part of this architectural scope. Technology remains very specific within this process and is treated as a realistic enabler of how Assembly designs and shares intention with stakeholders. This includes virtual and augmented reality, where Assembly's mobile applications serve to allow for more intuitive access to the projects to a broader range of stakeholders. Applications describe project aesthetics and materials as well as more "gateway functionality", allowing an ease of access to information. The more real-time this is, the better the feedback, which in turn allows the design team to better understand functionality within and beyond building construction. These operations expand agency while making the firm more competitive and safeguarding client satisfaction.

Still, geometry and formal part-to-whole relationships remain a large part of this visualization of data, and the firm has added value through the provisioning of other real-time datasets, including those describing schedule, cost, status, procurement, and entitlements. Each expand the discussion about, and impact of, architectural design.

NOTES

1 https://www.cimolai.com
2 The author was the project designer for this project/fabrication system while working at the firm from 1999 to 2002.
3 The Barclay's Center was a featured project in *BIM Design: Realizing the Creative Potential of Building Information Modeling* (Wiley, 2014), also by the author.
4 The Botswana Hub was extensively written about in *Workflows: Expanding the Architect's Territory in the Design and Delivery of Buildings* (Wiley, 2017), guest edited by the author.
5 S. Boschert and R. Rosen. "Digital Twin—The Simulation Aspect," in P. Hehenberger and D. Bradley (eds.), *Mechatronic Futures* (Springer, Cham, 2016), p. 61, https://doi.org/10.1007/978-3-319-32156-1_5.
6 See the section about the Barclay's Center in *BIM Design*.

IMAGES

pp 206–207 © GRO Architects; pp 209–212, 215–216, 217, 219–222 © Assembly OSM; pp 213, 217 © SHoP Architects

14 | A PRACTICAL SYNOPSIS

... A BRIEF HISTORY

No speculation on architecture in the twenty-first century would be complete without some notes on significant occurrences in the past. The cumulative understanding of these gave rise to the ideas put forth here, and have had consequences to the trajectory of architectural practice today.

Automation practices utilized in manufacturing today emerged from a twentieth-century goal of standardization, which promised efficiencies gained via a breakdown of a manufacturing process with workers repeating tasks within that sequence. This also led to a decrease in knowledge of the worker, giving rise to specialization, but more broadly impacting labor and the erosion of the nonlinguistically articulated knowledge held by craftspeople.

With Fordist practices commencing at the beginning of the twentieth century, knowledge became localized. Workers performed repetitive tasks along assembly lines, as in automobile factories. A worker in this case may, from the same position within that assembly line, put hundreds of cogs a day onto hundreds of car chassis without a larger understanding of how a *whole* automobile is put together.

This mode of operation is in direct contrast to a pre-modern way of making, with its foundation in the European guilds, where knowledge was conferred from an experienced member to an apprentice so that a holistic understanding of an overall process – an understanding of part-to-whole relationships – was gained. We need only to refer to DeLanda's blacksmith[1] in "Philosophies of Design" to understand this impact. The blacksmith received ore from one mine one week, another mine the next, and from a meteorite – an extraterrestrial source – the third week. All the ore was iron; however, its material properties were not homogenous, and some flexibility was required to work with the nuances of the material. This echo's Timothy Morton's call that architects should be more aware of our work materially – of its constituent *parts* as well as its *affects*. Craft implies a more holistic understanding of materials, and how those materials come together, both physically and procedurally, to form an assembly.

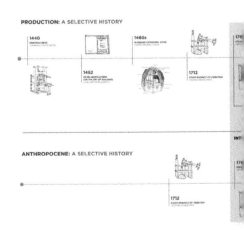

PRODUCTION: A SELECTIVE HISTORY

ANTHROPOCENE: A SELECTIVE HISTORY

This shift in labor practices was not a twentieth-century phenomenon. With Watt's steam engine, which came to maturity in the period between the 1760s and 1780s, the ushering in of more modern work practices had begun. The steam engine itself was invented, first and more crudely by Thomas Newcomen in 1712. For the approximately 50 years before Watt, Newcomen's steam engine effectively pumped water from coal mines, removing the interruption caused by flooding, common underground, in extracting coal. This corresponded with a drastic increase in the energy demands of humans, accomplished by the burning of coal.

Sanford Kwinter reminded us in 1996, "By the nineteenth century, the development of lithographic techniques allowed images to follow text into the public domain of mass reproducibility and mass circulation,"[2] which allowed for the codification of labor practices. Labor became subdivided into smaller chunks, making the sum of the parts less than the whole and the knowledge of individual workers became insufficient to achieve a fully formed product or process. Such a loss of knowledge would be fortified by Henry Ford, and his manufacturing process that would become known as Fordism, standardized for mass production and the consumption that allowed. Production, and *efficiency*, a main tenant of mass standardization, increased magnificently through this model.[3] This led to the broad dissemination of products and services, and to a larger extent, knowledge.

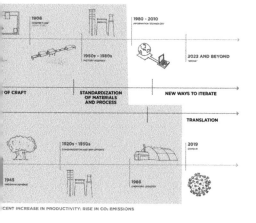

Figure 14.1 The printing press allowed information to be disseminated about, among other things, the practice of architecture as first codified by Alberti. Such a model of practice preferred a broader intellectualism to the trade specializations found in Gothic guilds. The loss of specialization, with other factors, ultimately led to a demise of the craftworker and the specialization of knowledge.

It is useful to understand this all with respect to architectural practice – itself codified through the mass dissemination of knowledge that came about through another vital invention, the printing press invented by the goldsmith Johannes Gutenberg in the 1430s. Alberti would introduce his De Re Aedificatoria less than 20 years later, in 1452. The codification of architecture largely happened as Alberti's text was one of the first architectural treatises to be broadly published, made possible by Gutenberg's invention.

Efficiency has been a guiding principle of labor practices and impacts the way we work and communicate. As of late, it has played a role in the relationship of the way humans have been depleting Earth's natural resources. In the simplest growth equation of our capitalist economic model, growth is never ending; and the last 300 years have seen fantastic increases in productivity and efficiency. The obvious issue is that Earth is a finite resource. It is massive though, and resources are distributed over space and time, preventing us from *seeing it* holistically, all at once. Still, we humans have been hard at work depleting it for our existence and well-being, but not always responsibly; sometimes, we have done it in a fantastical, magnificent, enormously destructive sort of way.

Figure 14.2 War-effort Assembly Line, Chrysler Plant, 1940s: During the twentieth century, as architects were engaging issues of interpretation of our design intent, many industrialists found a way to manage it as large-scale operations were given to assembly lines, where large groups of people stood in the exact same location and performed the exact same task for hours on end. These factories were retooled for the war effort in the middle of the twentieth century, augmenting capabilities and ultimately giving way to prefabricated construction products.

Taken through the twentieth century, following the war efforts of the 1920s and 1940s, with a historic economic depression in between, the Western world created a huge amount of factory space, and with it a large amount of knowledge pertaining to mass standardization and efficiency. These practices were applied to the manufacturing effort, which in the primarily non-wartime period of the second half of the twentieth century (Korea and Vietnam were *over yonder*) included many industries, including construction. Factories in the 1950s and 1960s began engaging concepts of prefabrication, and the rise of modular construction practices, especially with wood – an abundant commodity in North America – took hold. Precast reinforced concrete, an invention of chemistry and materials engineering, produced in controlled factory environments and trucked to project sites, also matured during this period.

Martin Heidegger's essays, "The Age of the World Picture" and "The Origin of the Work of Art," originally written in the 1930s, were published as this magnificent increase in productivity was being enjoyed across the Western world. Many positive aspects of this phenomena were felt in terms of information sharing and an increase in knowledge and innovation, which translated into both wealth and education levels. These increases in technology, and then productivity, would allow us to see the world in a different way.

Ultimately, efficiency was translated from these more industrial and machinic processes to those based in information technology (IT). Starting in the 1950s and 1960s especially in military research agencies,[4] IT through the application of computing led to further increases in business enterprise and engineering fields, with the latter crossing into medicine. Francis Fukuyama suggests that these increases and the end of the Cold War in the 1990s with the collapse of the Soviet Union are not coincidental.[5]

Figure 14.3 Modular Steel Systems, Bloomsburg, PA, 2020: As prefabrication began to take hold, some construction companies sought to relocate their operations to large climate-controlled spaces, where modular construction assemblies could be produced. The factory floor was reimagined to provide efficiencies in the assembly of construction products, but increased logistics by adding transportation requirements to building projects.

It was only a matter of time before these intensifications afforded through information technology began crossing into architectural production. Kwinter notes that information and experience merged at the end of the twentieth century.[6] This occurred in different ways with a broader implementation through the promise of efficiency. Advocates, many focused within schools of architecture, became more experimental in locating meaning through these technologies. Part of the issue of architectural production, especially as it relates to building information modeling (BIM), is that its adoption largely followed this efficiency schema. While we were certainly able to *make* buildings differently from both formal and relational perspectives – both with respect to components and the teams responsible for design and construction – the majority of practitioners were attracted to these technologies for reasons of efficiency. Architects could design buildings and implement those designs more quickly and comprehensively, which translated into an increase in design fees.

This is an important point if we assume there was not a significant *decrease* in the structure of architectural fees. In some instances, there was an increase due to the implementation cost of technology, so architects could earn the same amount of money for their scope of building design while completing it in a shorter time, again an increase in productivity. Computer-aided drafting (CAD), the precursor to BIM, had a similar impact, and slowly redistributed architectural labor. Designers were no longer tied to the drafting table, in a drafting room, but worked within a digital environment accessed via a computer terminal.

Still, the way we worked was consistent with predigital architectural production. CAD meant that designers were still drawing lines, which in the computer had properties that could be queried – such as length and curvature. However, they lacked the kind of object-intelligence that allowed a larger type of understanding of, and information sharing for, building knowledge. Drafting gave way to modeling, and modeling became more linked to database connectivity for data sharing and storage, itself requiring a rise of computing power that represented another fantastical increase in productivity and led to the more robust BIM systems we use today. Here labor takes a distinct turn. Younger designers, who were cheaper but less experienced in construction practices, brought fluency in computer programs. The digitalization of architectural information allowed for a new, and potentially more equitable distribution of knowledge within the design process. This shift affected pay scales within discipline and the very documentation required to convey design intent.

BIM as a process really allows for new ways of *seeing*, new ways of performing design activities that expand the territory and agency of the architect. BIM practically allows clash detection of building components in three-dimensional space, output of bills of materials, building costing, and generative design. It allows us to find those efficiencies, especially within the construction process. Broader simulation capacities allow architects to expand these themes to a more ecological perspective.

Through iteration, which includes the addition of specific material properties and the more precise understanding of metrics such as area, direction and volume, more responsive formal building proposal can be situated. Generative design is also tied to *automation*, both in terms of solution generation as well as of how parts can be produced and encompasses scripting and programming within its suite.

Thinking more ecologically, these operations can impact smaller or more local scales of design, including part-to-assembly relationships within a building, as well as more broad scales such as community or even region, allowing us to consider more holistically how our buildings are operated over time and interact with human and nonhuman phenomena.

But efficiency, as it is tied back to a capitalist model of production, where resources are understood as infinite, does not allow designers to really consider issues of depletion.[7] While this is not strictly a call for a steady-state, or zero-growth, economic model, it does speak to a reason to move away from efficiency as the sole basis for a digital-tool adoption in architecture while further increasing our awareness of how our work is interconnected to and affected by aspects of global climate change.

This interconnectedness also aligns architects with others who have primarily functioned outside of the discipline, such as eco-critics and environmental thinkers including Bruno Latour, Timothy Morton, and Páll Skúlason, each inspired in their own way by material logics or works of nature, such as Askja, an Icelandic volcano. Skúlason looked for ways in which humans could re-engage nature in both more productive and sympathetic ways.[8] Askja, and its environs, was one of the few areas left in the world where the idea of nature has not been impacted anthropocentrically, or redefined by human activity, hence NASA's interest in sending astronauts-in-training there in the mid-1960s.

NASA's work at this time allowed for the emergence of digital twinning, specifically in the operation and piloting of spacecraft within extraterrestrial environments. The high northern Icelandic plains so dear to Skúlason were the perfect location to create a twin, an analog to what NASA scientists imagined to be conditions that astronauts would face in space and on Earth's moon.

NASA's work near Askja focused on geologic exploration and collection, again made possible by the fact that those fantastical Icelandic plains were one of the only locales on Earth where nature had not been redefined by

Figure 14.4 Astronaut Monument, Húsavík, Iceland, dedication, 2015: The Astronaut Monument, was dedicated in 2015 by Neil Armstrong's grandchildren outside of the Exploration Museum, itself a dedication to human exploration, in Húsavík. It captures the names of 32 astronauts who underwent training around the volcano Askja between 1965 and 1967. Armstrong's name is prominently listed – second, under William Anders, the first human to photograph the earth, "Earthrise," from outside its atmosphere or territory, ushering in an objective way of seeing the world.

humans. Ultimately, 32 US astronauts traveled to Iceland for training between 1965 and 1967, and 14 went into space, with 7 of those performing geologic experiments on the moon's surface, which today remains the only extraterrestrial object visited by humans.

This training figured prominently in the Apollo programs, with William Anders, who commanded *Apollo 8,* being the first human to objectively photograph Earth, helping remove us from a subjective and terrestrial understanding of humans *being* in, or on the Earth to something that is outside of it, looking in. Earth, a hyperobject, after Morton, massively distributed in space and time became the focus of more objective consideration.

Páll Skúlason, the Icelandic philosopher, has noted how we need to newly understand our relationship to nature. Askja was important to him, as through it he links aspects of wholeness, or holism, to health, noting the words *whole* and *health* both originated from the same root in ancient Greece. Wholeness from an architectural point of view has been the basis of part-to-whole relationships in our work, which is to acknowledge that buildings are made up of a series of interrelated components, whether physical – stairs, walls, windows, mechanical units, etc. - or proportional. In a more subjective, and twentieth century, way of understanding this, parts were subjugated to the whole and were somehow less than it. In models of efficiency, this relationship is maintained in that the whole is greater than the sum of its parts.

The *interconnectedness* of things suggests that a disambiguation of these parts, especially as they relate to things both within and outside of a specific whole, should be considered, allowing for a more open ecological model to take hold. Griffiths and Kreisel use the spider web as a topography of encounter[9] that constitutes an open ecology, referring to it as a material engagement logic with both living and non-living things. Certainly, experiential ideas put forth by Morphosis and others, in both virtual and actual formats are timely here; as well as Zaha Hadid Architects' work with game engines as formal generators that drive ideas about *choice*, and how choice can be tested through building components, as well as environmental factors including the actions of a neighbor and those larger than a specific construction such as weather.

Open ecologies make a distinction between autopoietic and allopoietic objects and systems, suggesting the former has a deeper connection between energy and excess. Here Bataille's work on the general economy is relevant, and certainly his quantification of the sun as a source of renewable energy.[10] In this sense, a closed system, such as a factory, is really tied to a model of efficiency as it relates to human consumption, understood in Fordist images of the late-nineteenth and twentieth centuries and the fantastical productivity they set forth. A factory does exhaust resources in finite terms, while an autopoietic system has a more responsive understanding of part-to-whole relationships that seek not to deplete but cooperate in the use of resources, as they are engaged in the generation of new objects and experiences.

An open ecological model, perhaps the architectural project of the twenty-first century, can be accessed through enhancements in information technology in models of simulation that expand an understanding of building performance to suggest interconnectedness in the work of architects – buildings – to a broader environmental meshwork. Buildings become nodes in the mesh, points of intensity, where there is a higher level of definition, be it dimensionally, in a mathematical sense, or via computer graphics in the number of polygons used to specify a condition.

Conceptually, higher levels of definition convey design intent as well as ideas about *precision* in both technological and environmental ways, and how others translate our work in the actualization of things. Such digital working methods privilege point-line-surface development of NURBS geometry over the use of polygonal primitives. Polygons, albeit noneidetic ones via a literal thickening of NURBS geometry, form the workflow of virtual to actual constructions in that a polygonal translation of geometric form is achievable through automated, computer-numerically-controlled (CNC) material operations. Subdivision, or SubD, modeling, a popular type of modeling developed by Pixar Animation, is increasingly a modeling tool used by architects. Pixar maintains an open source library called OpenSubdiv and defines subdivision as "both an operation that can be applied to a polygonal mesh to refine it and a mathematical tool that defines the underlying smooth surface to which repeated subdivision of the mesh converges."[11]

Such a modeling workflow also defines a kind of part-to-whole relationship that predetermines componentry in the translation of a NURBS-based or SubD surface construction to a bespoke polygonal solid in an architectural organization. This idea is central in allowing architects to move from manufacturing paradigms that favor standardization to those of mass-customized architectural conditions. These acknowledge tool time as the basis of manufacturing efficiency given CNC hardware can produce multiple unique parts or the same part multiple times with equal effectiveness, applicable in both additive, or distributive, and subtractive manufacturing methods. It is notable that the same

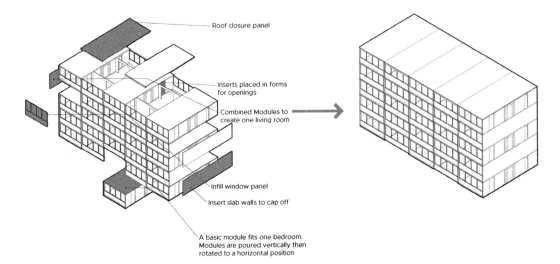

polygonal assemblage can be utilized in environmental operations, including shadow casting, solar gain, or more complex simulations involving energy consumption. In fact, this should become routine.

The mesh, then, as a translated assemblage of this process, becomes useful as both the basis for actualization of architectural componentry and as a digital twin aligned with new protocols in the design and production of an architecture that is open to new ways of *seeing* and understanding this twenty-first century world.

NOTES

1 DeLanda, Manuel, "Philosophies of Design, the Case for Modeling Software," in *Verb Processing: Architecture Boogazine* (Actar, 2001), ISBN: 9788495273550.
2 Sanford Kwinter, "Virtual City, or the Wiring and Waning of the World," *Assemblage* 29 (April 1996): 86–101.
3 Richard Garber, "Digital Workflows and the Expanded Territory of the Architect," *Workflows: Expanding Architecture's Territory in the Design and Delivery of Buildings* (Hoboken, NJ: Wiley, 2017), ISBN: 9781119317845.
4 Branden Hookway, *Pandemonium: The Rise of Predatory Locales in the Postwar World* (Princeton Architectural Press, 1999).
5 Fukuyama Francis, *Our Posthuman Future: Consequences of the Biotechnology Revolution.* 1st Picador ed. New York, Picador, 2002.
6 Kwinter, pp. 86–101.
7 Herman Daley, *Steady-State Economics*, 2nd ed. (Washington, DC: Island Press, 1991), ISBN 978-1559630719.

Figure 14.5 Shelly Systems Precast Concrete Modular System by GRO Architects, 2023: Operation Breakthrough was a 1970's era demonstration project by the federal Department of Housing and Urban Development (HUD) that tested innovative building materials and construction methods with the goal of removing obstacles to large-scale affordable housing production in the United States, bringing quality housing to all income groups. Operation Breakthrough ultimately worked with 22 "system producers" that provided some 2,900 housing units, all delivered using off-site construction methods. The selected producers utilized housing systems ranging from precast concrete- or wood-framed modules to units constructed largely of plastic or metal, supplied by companies including Alcoa, Levitt Technologies, General Electric, and Republic Steel. The housing was installed in selected urban markets, including Jersey City, NJ which was the only northeastern demonstration site. There, Shelly Systems, a then Manhattan-based modular developer, produced a precast concrete modular unit executed in low-, mid-, and high-rise configurations, bringing about 500 affordable units to the city in 1975.[12]

8 Páll Skúlason, "On the Spiritual Understanding of Nature," lecture given at Ohio Northern University, April 15, 2008.
9 Devin Griffiths and Deanna Kreisel, "Introduction: Open Ecologies," in *Victorian Literature and Culture* 48 (1): 1–28, © Cambridge University Press, 2020, p. 3.
10 Georges Bataille, *The Accursed Share: An Essay on General Economy* (Zone Books, 1988).
11 See https://graphics.pixar.com/opensubdiv/docs/subdivision_surfaces.html#piecewise-parametric-surfaces, accessed October 22, 2022.
12 Shelly Systems diagram originally produced by Shaohong Tian in the author's fall 2022 Urban Design Studio at Penn.

IMAGES
pp 224–225, 227 © GRO Architects; pp 226 © Bettmann Archives, Getty Images; pp 229 © David Philips / The Exploration Museum

INDEX

2D vector information, providing, 188
3D building geometry, 188
3D concrete printers, franchising opportunities, 160
3D modeling, alternative, 10f–11f
3D printers/printing, 159–161
3DStudio MAX (polygon-based modeling program), 127
3D Studio, usage, 218
4IR. *See* Industry 4.0 movement

Abnatural open ecological approach, 129
Abstraction, vagueness (contrast), 7
Academy Street study (GRO Architects), 222f–223f
Achi-Tectonics, manufacturing engagement (history), 133–134
Acrylonitrile butadiene styrene (ABS) plastic, usage, 159, 160
Actualization, 161–162
Adaptive models, creation, 63–64
Affordance matrix, 218f–219f
"Age of the World Picture, The" (Heidegger), 226
Alberti, Leon Battisti, 49, 84, 170–171, 225
Alexander, Christopher, 17
AlWasl Tower (UNStudio), 62f–63f
Amaterial virtual environments, 166–172
American Concrete Institute (ACI), 191
American Institute of Architects (AIA)
    Climate Action Plan, 122–123, 124f
    mitigation, 123
Americans with Disabilities Act (ADA), 186, 188, 202
Amsterdam Central Station, 69
Anders, William, 230
Anthropocene, 82, 135
Apollo
    astronaut training (Iceland), 87f
    Command Module Mission Simulator, 52f, 53f
Aquifers, thermal mass usage, 129–130
Aquifer Thermal Energy Storage (AETS) strategy, 77–78

Archipelagos, creation, 139
Architects
    developers, relationship, 221
    work, commoditization, 221
Architectural geometry, repackaging, 97
Architectural model, precision, 48
Architectural practice, codification, 19
Architectural precast concrete, usage, 188–189
Architectural project, projective optimism, 136
Architecture, Engineering, and Construction (AEC), 214
    disciplines, 55
    industry, changes, 84–85
Architecture, observation, 127–130
Arnhem Train Station, UNStudio work, 61–62, 63
Artificial intelligence (AI), 101–102
    integration, 64–65
Ash Meadows Fish Conservation Facility, 91f, 92
Asian Games Park (Architectonics), 132f–133f, 136–138
    archipelagos, creation, 139
    assemblage, synthetic nature, 139
    both-and-condition, 145
    building, 145, 147
    cong, inspiration, 144f
    construction, initiation, 149o
    COVID pandemic construction, occurrence, 140
    cranes, deployment, 141f
    design intent, ensuring (virtual means), 149f
    design process (Dubbeldam), 137
    drawing set, 141f
    Dubbeldam team/BIM consultants (collaboration), 141f
    earth-building, 138–143
    ecological strategies, 139–140, 139f
    façade glazing system, variable (triangular) frame (extrusion), 147f
    field hockey stadium, 144f

Asian Games Park (Architectonics)
(*continued*)
  green shopping valley, creation, 138
  human users, 140
  information sharing, extensiveness, 140, 143
  lessons, learning, 137
  nature, conservation, 136
  non-human users, 140
  object with/in strategy, usage, 146f
  park structures, 136–137
  planar glazing panels, usage, 148
  post-games use, 139–140, 139f
  roof, large-span structure solution, 145f
  site, 137–138, 137f
  stadia, 143–149
  structural roof solution, 147–148
  suspendome, usage, 146f, 147–148
  table tennis stadium
    entry sequence, 145f
    nesting strategy, 150f–151f
    suspendome, usage, 146f
  three-dimensional models, usage, 140
  topography, variation, 142f–143f
  white elephant status, avoidance, 146f
  workflow, usage, 148
  zero-earth strategy, 138–139
Askja (Iceland), 83f, 229
Assembly, 172–174
  approach, 211–212
  building futures, 206–212
  floor-ceiling assemblies, 181
  model, usage, 210
  natural progression, 210–211
  platform, 214–215
  proof-of-concept product, existence, 213–214
  resistance, relationship, 181–183
  system, function, 201
  value, 214
Assembly line, 11f
  mass-customized assembly line, 12f
Assembly line (Chrysler Plant, 1940s), 226f
Assembly OSM, 198
  chassis assembly, 209f
  materials, usage, 203f
  mechanical module, 211f
  modular assembly limitations, 204f
  module interior, 213f, 214f
  module kitchen, 213f
  module production, 201f, 202f
  process automation, 207f
  prototype stacking, 208f
  space race, 204
ASTM International, mixes (concrete), 192
Astronaut Monument (Húsavik), 229f
Astronaut training (Iceland), 87f
Atriums (atria)
  external inputs/internal criteria, awareness, 31
  green lung function, 78f
  social mixer function, 30, 41f, 42f
Augmented Field of Architectural Design (GRO Architects), 218f–219f
Augmented reality (AR) tools, 208
AutoCAD, usage, 214
Autodesk Maya, usage, 113–114
AutoLISP, 218
Automated feedback, 8
Automated Range of Solutions for a Site (GRO Architects), 13f
Automated techniques/workflows, process (impact), 9
Automation, 216–220
  practices, usage, 224
Automation, usage, 65
Autopoeitic/allopoetic objects/systems, distinction, 230–231
Autoring tools, usage, 215
A-Wall (one-to-one scale exercises), 203

Bauldrillard, Jean, 92–93
Bevier, Doug, 190
Bhooshan, Shajay, 96, 99, 101, 102
  computer graphics, interest, 103–104
  de-risking, meta-layer, 115–116
  solutions, user activation, 105
  tectonics, alignment, 112
  three-dimensional configurator environment navigation, 109–110
Bidirectional lifecycles, 66–69
BIM Capability and Assessment (GRO Architects), 14f–15f

BIM Design: Realising the Creative Potential of Building Information Modeling (Garber), 24
BIM Design techniques, advancement, 162f–163f
Blanchard, Gaston, 48
Bloch, Ernst, 127
Boeing Offsite Manufacturing, assembly OSM (usage), 198f–199f
Bogost, Ian, 19–20
Booking.com headquarters (UNStudio), 58f–59f, 70f–74f
  atria, green lung function, 78f
  balconies/amenities, human scale, 76
  BIM, uploading, 80f
  boundary conditions, 72–73
  campus, 69–74
    collective massing strategy, 128
    housing component, 73–74
  double-height plinth, 76–77
  floorplates, flexibility, 79f
  formal transitions, allowance, 73
  glazing strategy, usage, 73
  humans/services, circulation, 74–78
  lifecycle concepts, 79–80
  massing, floorplates (depth), 75f
  mechanical strategy, development, 78f
  public circulation, 74, 76
  public interface, 74f
  textures, 75f
  unitized glazing system, 76f–77f
Brainport Smart District project (UNStudio), 65
"Brittleness and Bureaucracy" (Spencer), 49
Brown, Peter Hendee, 221
BSA approval, 204
Builder, The (program), 17
Building as Material Banks (BAMB), 66f
Building information modeling (BIM), 63
  adoption, 60
  client understanding, 18
  connections, 64
  design, 84
  framework, protocols (establishment), 14f–15f
  future, 56
  impact, 47
  platform
    precast systems, plug-in functionality, 189
    scalability, 66
  process, 228
  seeing architecture, contrast, 120f–121f
  software, usage, 6
  systems, 52, 208
  thinking, application, 223
  usage, 45, 79, 189
  workflows, setup, 67–68
Building information modeling (BIM) technology, 29, 53, 114
  adoption, 7
  impact, 103–104
  position, 8f–9f
  usage, 118–119
Buildings
  actualization, abstraction (power), 7
  code requirements, 191
  delivery, paradigm, 67
  greenhouse gas contributors, 84
  intelligence, 203–204
  issues, 158–159
  systems, density, 176
Business as usual approach, 57

CAD/CAM
  production, 55
  revolution, 53
Capital flows, technological workflows (relationship), 12–15
Capital-forward motives, 222
Carbon dioxide emissions, reduction, 64
Carbon dioxide levels, 82–83
Carbon emissions, 191–192
Cartesian dimensions, errors, 159
Cartesian space, larger-dimensional increments, 160
CATIA, 12, 24, 44, 100, 218
  3D environment, 198
  development, 28
  digital linkage, creation, 35
  environment, 25f, 37, 201, 207f
  geometric affect, incorporation, 31
  geometry, 37f, 40
  guide panel adjustment, 43f
  platform, 202f, 203, 205

CATIA (*continued*)
  polysurfaces, usage, 32
  technology, 200
  usage, 43f
  workflow, 205
Cement production, carbon dioxide
  emissions, 84
Cerone, John, 54, 206–209, 213–214
Certification, offering, 19
Chance encounter, 31
Choice/consequence, 31–32
Cimola (chassis supplier), 200
Cinema4D (polygon-based modeling
  program), 127
Circular economy (Hadid), 94
Circular factory, ZHA role, 98
Clark Pacific (concrete company),
  189–190, 192
Climate Action Plan (AIA), 122–123, 124f
Climate Action Tracker, 122f
Climate change, deceleration, 122
Climate Imperative, 122–123
CODE group (ZHA), 96–97
Collective massing strategy, 128
Communication capacities, increase, 35
Competitive bidding, 20
Components, standardization, 182
Componibile, 166
Compositional-class objects,
  polymorphism (achievement), 16–17
Compositional feedback, 96–97
Compositional open ecological
  approach, 129
Computer-aided drafting (CAD),
  103–104, 228
  systems, usage, 6, 8
Computer-numerically controlled (CNC)
  capabilities, 192
  flatbed routers, usage, 159
  hardware, 55, 216, 231–232
  machines, usage, 220
  material operations, 231
  models/parts machining, 161
Computer, usage (justification tool), 11
Concrete, cement content, 191–192
Concrete masonry unit (CMU) block, usage,
  186–187

Configuration-to-Delivery (SHoP
  Architects), 205f
Configurator (ZHA)
  deployment, 99
  development, 98
  sessions, running, 108–109
  usage, 101–102
Cong (object), 143, 144f
Construction
  considerations, 44–45, 154, 178
  means/methods, 157–159
  modular construction, 178
  permits, obtaining, 185
  process, meta-control, 64
  timescales, 156
COVID-19, production
  challenges, 182–183
Co-virtual reality (VR) space, usage, 54
Craft, implication, 224
Craftsmanship, tradition, 175
Crafts-work, 174
Creative thinking, usage, 221
Crutzen, Paul, 82
Curvilinear geometry, usage, 18
Customization
  goals, misalignment, 154
  programming, impact, 159
Cylindrical stratum, planar material
  (addition), 39f

Dalux, 79–80
Daly, Herman, 88
Darwin, Charles, 125
Dassault Systèmes, 28
Database, creation, 67
Data, leveraging, 66–67
Dave, Bhargav, 157
Dead projects, 32
"Dead" three-dimensional models, 55
Dean Street project (SHoP), 198
Decentralized platform model, usage, 54
DeLanda, Manuel, 48, 64, 101–102,
  127, 134f, 224
Department of Building (DOB)
  standards, 202–203
De Re Aedificatoria (Alberti), 84,
  170–171, 225

De-risking, meta-layer, 115–116
Design
    content, gamification, 55
    documentation, completion, 67
    information, 155
    inside-out approach, 73f
    intent, creation, 158
    iteration, interdependency (impact), 157
    logics, processing, 207
    relationships/inputs, importance, 171
    tools, codification, 157
    workflows, 171
Design-assist, impact, 190
Design-bid-build process, streamlining, 36
Design/construction
    exercise (GRO Architects), 186–187
    space, collapse, 47
Design Patterns: Elements of Reusable Object-Oriented Software, 17
Design process, meta-control, 64
Design surface, 25f
    two-dimensional/three-dimensional location, 36–37
Design-to-build team, expansion, 206
Design-to-production schemas, 192, 196
Deutsch, Randy, 56–57
Development space, 220–223
Devil's Hole
    pupfish, presence, 89–91, 90f
    simulacrum, 92
Diagonal jointing, 38f
Diagonal joint pattern, usage, 39
Digital data, production (CATIA usage), 203
Digital design/construction, benefits, 9
Digital design operations, impact, 211
Digital design, technologies (maturation), 118–119
Digital infrastructure, 43f, 44
Digital linkage, creation, 35
Digital object, interactions, 18
Digital Project (Gehry Technologies), 218
Digital script, development, 40
Digital technologies, adoption, 206
Digital tools
    simulation capabilities, 123
    usage, 20, 27

Digital twin
    assembly usage, 209–210
    concept/environment, 51–55, 205
    data/systems series, 208
    present object, 56
Digital Twin, 32, 198
Digital twinning, 18
Direct-to-manufacture interface, 160
Disaggregation/separation, 125
Di Stefano, Nicola, 88
Double-height plinth, usage, 76–77
Drafting, impact, 228
Dubbeldam, Winka, 135–136
    design process, 137
    flatness, absence, 143
    para-building, 134
    synthetic natures, 132

Earth-building, 138–140
Earthrise, 119f
Eco-criticism, 136
Ecological crisis, 125
Ecological thinking, pervasiveness, 119
Ecological Thought, The (Morton), 125
Ecological twin, 89–93
Ecology
    considerations, 82, 118
    vastness, 83–86
Economies of scale, discovery, 182
Economy/ecosystem, throughput, 89f
Ecosystem, thriving, 92
Efficiency/flexibility, codification, 63–64
Efficiency, principle, 225–226
Efficiency to Seeing (GRO Architects), 120f–121f
End-to-end manufacturing processes, 156f–157f
Environmental performance simulation (UNStudio), 128f
Environmental plug-ins, 135f
Errors and omissions (EO) insurance, usage, 158
Evens, Aden, 18
Extended reality (XR), 27, 29
Exterior envelope model, release, 37

Façade glazing system, variable (triangular) frame (extrusion), 147f
FaceTime, usage, 167
Factory assembly, meaning, 181
Factory logics, 184
FEMA standards, 205
Field hockey stadium (Asian Games Park), 144f
Financial market analysis (FMA), 198
Finite element analysis (FEA), usage, 189
Flexible models, usage, 24–28
Floor-ceiling assemblies, 181
Floorplates (UNStudio), 74, 79f
Floridi, Luciano, 50–51
Fly ash, usage, 192
Ford, Henry, 225
Fordism, 225
Form-giving, absence, 65
Formwork
    assembly/reusability, understanding, 192
    reusability, 160–161
Form-Z, 218
Four Historical Definitions of Architecture (Parcell), 9
Frankfurt Four (UNStudio), 161f
Frog Design, 200
Full Cycle of Life (GRO Architects), 17f
Full-scale concrete printing services, launch, 160–161
Furniture
    components, 173
    production, Archduke Ferdinand (impact), 174
Furring channels, usage, 45
Future-oriented hope, 127

Gantt chart, 14f
Gartner Hype Cycle for Emerging Technologies, 8f–9f
General Services Administration (GSA), embassy design (SHoP collaboration), 205
Geologic time, term (usage), 86–87
Geometric basis, 163–164
Geometric processes, connections, 52
Geometric response, 221
Geometry, metadata (association), 31–32
Geopolymer concrete, defining, 192
Ghost Kitchen trend, 53–54
Glass window-wall, usage, 175f
Glazing strategy, usage, 73
Global thinking, 25
Google Chrome, usage, 113
"Gothic Architecture by Remote Control" (Toker), 170–171
Graphic user interfaces (GUIs), usage (example), 13f
Greenhouse gas emissions, impacts, 122–123
Green shopping valley, creation (Archi-Tectonics), 138
Green spaces, increase (support), 135–136
Greenwich Street Building (Architects), 133–134, 134f
Griffiths, Devin, 125, 127, 129–130, 230
Ground granulated blast furnace slag, usage, 191
Growth
    limitations, 226
    resource-intensive capitalist model, 130
Gutenberg, James, 225
Gypsum board, absence, 202–203

Hadid, Zaha (circular economy), 94
Haeckel, Edwin, 125
Harraway, Donna, 119
Harris, Kerenza, 24, 29–32, 34, 44
Holocene, 82
Homo-mensural principle, 88
Hoppermann, Marc, 58, 61, 67, 69, 76
Horizontal inputs, allowance, 64
Horizontal ribbing, concept, 45
Human-computer interfaces (HCIs), 65, 154, 216
Human resources management, 155
Human-robot cooperation, 220
Humans/services, circulation, 74–78
Hurricanes Harvey/Ian, impact, 86
Hyperobject, term (usage), 86

I-joists, usage, 179
Immaterial data flow, 56
Immersive environments, 24–28
Immersive reality research, design process (connection), 31–32

Industry 4.0 movement (4IR), 53, 213–214
Info-realism, 48–51
Information
    modeling, usage, 185
    objects, 6–7
    sharing, 140, 143
    transfer, 183–185
Informational realism, 50–51
Information modeling, 86, 89
Information technology (IT)
    appearance, 118
    process basis, 227
    usage, 154–155
Infrastructure, understanding, 85
In-house labor, usage, 183
Inspection, impact, 185–186
Integration platform model, usage, 99
Interactive models, creation, 63–64
Interconnectedness, 17, 127–128, 229
Interconnectedness of things, 230
International Building Code (IBC) Type V-A construction, 186–187
International Code Council (ICC), 191
Internet of things (IoT), usage, 53
Interoperability, 44
    allowance, 55–56
Interstitial space, 176
Italian manufacturing, craftsmanship (tradition), 174

Joint pattern, application, 43f
Jones Hall (GRO Architects), 50f–51f, 159f
Journal Square Redevelopment Plan, 172–173
Just-in-time inventory, 155

Kaleidoscope VR, 28
Kalsaas, Bo Terje, 157
Katerra
    buildings, 155f
    experimentation, 155
    goals, 154
Kolbert, Elizabeth, 82, 89, 92–93
Koskela, Lauri, 155
Kreisel, Deanna, 125, 127, 129–130, 230
Kwinter, Sanford, 225

Large-scale physical models, building, 207
Large-span structural solution, 145f
Latour, Bruno, 62–63, 122, 229
Leach, Neil, 16–18, 218
Lean
    components, 155
    construction, 156–157, 158f–159f
    principles, 157
    processes, development, 157
Liberland (ZHA), 115f
LIDAR/3D scanning/photogrammetry, 210
Lifecycle concepts (Booking.com), 79–80
Lifecycle phases, 51–52
Light-gauge metal studs, usage, 45
Light-gauge steel construction solution, 186–187
LISt Programming, 218
Lopez, German, 122
Loss factor, 222
Lynn, Greg, 16

Machine code, generation, 220
Machine learning, usage, 65
Madrazo, Leandro, 16
Manufactured wood-modular systems, site-built wood frame structures (similarity), 179
Manyfesto (Griffiths/Kreisel), 129–130
Mass-customized assembly line, 12f
Mass-customized generative components, 134
Massing, 138
    collective massing strategy, 128
    surface extraction, 44
Mass-production, 154–156
Mass-standardization, paradigm, 220
Mass-standardized assembly, 10f
Material
    assemblages, 171
    constructs, workflows (exposure), 28–31
    standardization, 182
    thickening operation, 37, 39
Material passport, 60–61, 68
Material systems, design, 85–86
Maturana, Humberto, 130
Mayne, Thom, 24, 27, 31
McDermott, Patrick, 156
McDowell, Alex, 29
MDF, usage, 203

Mead, Rebecca, 84
Means and methods, 214–215
Mechanical, electrical, and plumbing (MEP)
   design teams, impact, 111
   scope, 183
Mechanical strategy, development, 78f
Mechanical sympathy, 46–48
Mercedes-Benz Museum (UNStudio), 63
Mesh
   functions, 126–127
   interconnectedness, 126
   polygon meshes, usage, 127
   usefulness, 232
Meta-control, 64
Metaprocess, 29
Metaverse, workflow (gaming), 102–114
Micro-factory, 106, 1086
Microsoft Edge, usage, 113
MicroStation (Bentley Systems), usage, 32, 35, 37f
Model-based environment, immersion, 55
Modeling
   design process, 30
   progression, 218
   technologies, usage, 88–89
   tools, schema (trends), 7, 20
Modular building, 178, 182–183
Modular company, dimensional capacity, 184
Modular construction, 178
Modular Fabrication, information basis (usage), 184f
Modular kitchen (Snaidero "Spazio Vivo"), 173
Modular materials, workflow (relationship), 179–181
Modular platforms, 198
Modular production basis, contemplation, 154
Modular Steel Systems, 186, 227f
Modular systems, vertical continuity (absence), 185–186
Modular thinking, understanding, 175
Modular unit-boxes, fitting, 183
Modular units, shipping, 182
Montgomery Mosque (GRO Architects), BIM tools (usage), 14f–15f

Morphosis
   physio-spatial/nonphysical strategies, 30
   workflow, 25
Morton, Timothy, 86, 125, 224, 229
   holism, 125–126
   interconnectedness, 126
   meadow, 128
Moxie (GRO Architects)
   construction documentation, re-engagement, 187–188
   modular building configurations, 182f
   modular dwelling unit, 179f, 181f
   modular Moxie, 186–188
   PTACs, requirement, 187f
Muchamp, Herbert, 134
MudBots, in-house solutions, 160–161
Multi-axis routers, usage, 192

Naming conventions, application, 44
Natural materials, usage, 92–93
Natural pozzolans, usage, 191
Natural, term (consideration), 86–89
Nature, intelligence (learning), 135–136
NBIMS, transfer protocol debate, 58, 60
Nest Micro-Housing (GRO Architects), 166f–167f
   building
      density, 168f–170f
      height, permission, 173
   built-in cabinetry, 174
   characteristics, 167–168
   demising walls, 177f
   dry zone, 172f
   front façade, 176f
   furnished dwelling units, 175f
   glass window-wall, usage, 175f
   panelization, 178f
   spatial transformation, allowance, 173f
   unit, design, 172
   wet zone, 177f
   window boxes, projection, 176f
Newcomen, Thomas, 82, 224–225
Next Generation Precast Concrete, usage, 194f–195f
NFTisms (ZHA), 116f
NOAA Sea Rise Viewer, 85

Nodal Development (GRO Architects), 127f
Nodal logic, 126–127
Non-bodies, harmony (absence), 88–89
Noncomputational workflow, limits, 48–49
Nonprogrammatic open ecological approach, 129
Northeast Precast (production), 189f, 192
NURBs
    curve, manipulation, 56
    geometry, 127
    modelers, 218
    point-line-surface development, 231

Object
    confrontation, 89
    with/in strategy, usage, 146f
ObjectARX, 218
Objecthood, implication, 17
Objective mediation, 48
Objective thinking, shift, 15–18
Object-oriented programming (OOP), 16–17
    shift, 51
Object-oriented virtual modeling, 49
Object-precision, 48–51
Off-site building, 178
Omniverse (Nvidia), 56
One-to-one process, 207
On-site stacking, 198
Open ecologies, 129–130
    distinction, 230–231
Open Ecologies, 125
Open-frame wood trusses, usage, 179
OpenSubdiv, 231
Operational carbon, measure, 191
Operations, impact, 47–48
Optimization
    finite element analysis (FEA), usage, 189
    result, 184
Orange County Museum of Art, 33–43
    atrium cladding, 41f, 42f
    cladding optimization, 43f
    cylindrical stratum, planar material (addition), 39f
    design geometry, change, 43f
    diagonal jointing, 38f
    envelope geometry, studies, 35f
    façade, 25f, 34f
        local texturing, 39–40
        surfaces (population), rule-based system (usage), 42
    geometry, importation, 37f
    guide panel adjustment, 43f
    immersive environment, 27
    pedestrian access/plaza activation, 33f
    ribbon geometry, 36f, 39f
    ribbon, tiling pattern (both/and strategy), 40f
    surface tiling, geometric solutions, 38f
    tiling process, 40
Origin of Species (Darwin), 125
"Origin of the Work of Art, The" (Heidegger), 226

Para-building, 134
Parcell, Steven, 9
Part-to-whole project relationships, 7f
Part-to-whole relationships, 16–17, 119, 125, 127–128, 129f, 174
    understanding, 231
    wholeness, basis, 230
Part-to-whole understanding, 155
Pattern Language, A (Alexander), 17f
Pattern, term (usage), 17
Pedestrian bridges (!melk), 138
Performance modeling, 87
Perimeter drain, providing, 87
Permitting
    achievement, 185
    usage, 185–186
Perot Museum of Science and Nature, completion, 24
Phare Tower (Lighthouse Tower), 24, 26f, 27
"Philosophies of Design" (DeLanda), 224
Physical-by-virtual operations, 210
Physio-spatial/nonphysical strategies, Morphosis usage, 30
Pinnell, Renee, 28
Pixel organization, resident-participant selection, 111f
Planar glazing panels, usage, 148
Plasma cutters, usage, 192
Platform thinking, importance, 99

Point-line-surface workflow, introduction, 126
Polygonal meshes, usage, 113, 231
Polylactic acid (PLA), usage, 159
Polymorphism, achievement, 16–17
Polysurfaces, usage, 32
Portland Cement, usage, 188
Power, distributions (unevenness), 129
Practice, considerations, 216, 224
Precast
  concrete, usage, 188
  design-to-production schemas, 192, 196
  emissions, 191–192
  industry, standardization, 189–190
  systems, plug-in functionality, 189
Precast, popularity, 178
Predesign, 13
Pre-engagement services, 13
Prefabricated components, ZHA creation, 97–98
Prefabrication schemes, 200
Pretensioned panel designs, exploration, 190
Prix, Wolf, 54
Pro-active productive system, 135
Problem-solving process, 96
Procedural Modeling Basis, 10f–11f
Process, industrialization, 212–214
Process modeling, 8–9
Production, automated basis, 163–164
Project aesthetics, discussion, 190
PTACs, requirement, 187f
Public circulation, 74, 76

Quadrilateral polygons, usage, 126–127
Quotas, usage, 88

Range, notion, 127
real, idea (initiation), 50
Rebuilt surfaces, trimming, 35
Re-engineering process, 25
Reinforced concrete, usage, 188
Remote control, 166, 169
Remoteness, 174–177
Repeated module, importance, 171–172
Research-based endeavor, 96
Research, notion (Heidegger), 124
Residential bar typology, 184

Resource prices, efficiency (requirement), 88
Reverse engineering, 56
Revit (Autodesk), 12f, 190, 218
  Autodesk Revit environment, 56, 140, 188, 208
  interface, 32
  platform, 189
  workflow, 205
Revit>CATIA workflow, establishment, 205–206
RFI procedure, 214–215
Rhinoceros 3D, application, 56, 135f
  engineering coordination, 113
  interface, 32
  polysurfaces, 32
  3D, 56, 135f
  usage, 43f, 44, 145, 218
Ribbon geometry, usage, 36f
Rivera, Daniel, 156f
Roatán Próspera Residences (ZHA), 94f–95f, 100–102, 103f
  aesthetics, 102
  buyer choices, capture, 110f
  computational techniques, 105
  configuration, 105f
  developer space, 102
  housing typologies, issues, 111
  imagining, modular fabrication/assembly, 114f
  individual user input, visualization, 108
  intuitive manipulations, 109f
  load, forming/loading, 104–105
  modular housing proposal, 104–105
  online interface, 106f–108f
  pixel organization, resident-participant selection, 111f
  production/workflow, targeted/sustainable aspect, 101
  revised floor plans, circulation, 112
  typologies, standardization, 113f
  units, bookending, 112f
  user-based geometries, 112–113
  user input, receipt, 110f
Robinson, Kim Stanley, 119
Robot arms, usage, 159, 160

Robotic and Additive Manufacturing Lab (Weitzman School of Design), 160
Rubenstein Commons (Steven Holl Architects), 193f
Rule-based system, usage, 42
Russo, Maria Teresa, 88

Scalability, meaning, 113
Scaled model, usage, 206
Schumacher, Patrik, 112
Scripting, usage, 13f, 218
Sea Level Change Science Team (NASA), 85
Sea Rise Viewer (NOAA), 85f
Self-maintenance cells, involvement, 130
Serra, Richard, 33, 34f
Sharing, idea, 140
Sharp, Alex, 156
Sharples, Christopher, 201
Sharples, William, 201, 203
SHoP Architects
    co-VR space (Tranquility), 54f
    design research, 198
    digital transfer, usage, 203
    project delivery success, 206–207
Silica fume, usage, 192
Simulation
    achievement, 51–52
    technologies, usage, 88–89
Simulation and Its Discontents (Turkle), 46
Site-construction teams, usage, 185
Skúlason, Páll, 83, 229, 230
Smart components, prefabrication requirement, 134
Smart workplaces, usage, 65
Snaidero "Spazio Vivo" modular kitchen, 173
Softimage (polygon-based modeling program), 127, 218
Sole-sourced products, 201
Sound transmission class (STC) ratings, 190–191
Source of truth, 207
Spaces
    loss factor, 222
    race, 204
Specialization, loss, 225f
Spencer, Matt, 49
Squint Opera (UNStudio), 68f

Standardization, 183
    practices, manifestation, 210
    rules, 180
Steady-state economic model, 88
Steel chassis, flat packing, 200
Stick-built site construction, similarities, 185
St. Nick's (SHoP Architects), 209f
Straitus Bridge (ZHA), 104f, 112
Structural concrete, building code requirements, 191
Structural precast, 189
Subdivision (SubD) modeling, 56, 231
Sugiuchi, Atsushi, 25, 29, 31
Superior Walls®, usage, 183
Superusers (Deutsch), 56–57
Supply chain
    creation, 200
    engagement, 212
    logistics, 68
Surface/volume (defining), mesh (usage), 126–127
Suspendome, usage (Asian Games Park), 146f, 147–148
Sustainability, ZHA approach, 104
Sympathy
    impact, 46
    term, resonance, 48
Systems-level thinking, increase, 125
Systems, total value, 190

Techné, concept (elaboration), 9
Techniplan Adviseus, UNStudio (interaction), 77
Technique > Process > Workflow, 161–164
Technological models, automated ranges, 11
Technological ontology, 8–9
Technological workflows, capital flows (relationship), 12–15
Technologies of consensus, 112–113
Technology
    improvements, 171
    stack, 103–104
Tectonics, alignment, 112
Tekla, 12f, 189
    platform, 189
    usage, 190
Temporal relationships, 206

Tender drawings, codification, 206
Third-party subcontractors, hiring, 183
Three-dimensional data-driven environment, impact, 20
Three-dimensional data, translation, 25
Three-dimensional geometry, rationalization, 196f
Three-dimensional human-scale basis, 86
Three-dimensional mesh, usage, 37
Three-dimensional modeling software, usage, 16, 27f
Three-dimensional model, providing, 6, 15f
Three-dimensional process, usage, 30
Throughput, 89f
Tier 1 assembly space, 204
Tier 1 system, 202–206
Tier 2 companies, components, 202
Tier 2 fabricators, building part manufacture, 201
Toker, Franklin, 48, 170–171
Tomasetti, Thornton, 133f, 136
Tool-authoring, 19–20
Topography, variation (Asian Games Park), 142f–143f
Total productive maintenance, 155
Total quality management, 155
Toyota Production System (TPS), 155
Traditional Modular Workflow (GRO Architects), 180f
Transfer protocol, debate, 58, 60
Triangular polygons, usage, 126–127
Turkle, Sherry, 46, 192
Two-dimensional CAD systems/packages, usage, 6, 216
Two-dimensional capacity (drawing production), 185
Two-dimensional drawings
  interpretation, removal, 214–215
  usage, 55, 212, 223
Two-dimensional plane, 108
Two-dimensional tender sets, 42
Two-dimensional/three-dimensional workflows, 52
Two-dimensional unfolded surfaces, 44
Typology, 99, 156
  ideas, evolution, 16

Under a White Sky (Kolbert), 89
Undercutting, usage, 159
Underground Thermal Energy Storage (UTES), 128–129, 130f
Unitized glazing system, 76f–77f
Unity (interface), 56
UnReal Engine (interface), 56
UnReal Engine, usage, 113
UNStudio
  BIM Transition, 60f–61f
  daylighting/circulation simulations, 79–80
  environmental performance simulation, 128f
  floorplates, usage, 74
  future lifecycles, 58
  Techniplan Adviseurs, interaction, 77
UNStudio, usage, 53
Urban massing, 72f
User-based geometries, 112–113

Value adding equation, 220–221
Value-building, 64
Van Berkel, Ben, 61, 62, 64
Varela, Francisco, 130
Vastness, 83–86
VectorWorks, 218
Vertical continuity, absence, 185–186
Virtual>Actual, (BIM Design usage), 49f–50f
Virtual counterpart, testing/simulation process, 205
Virtual Design and Construction, technology position, 8f–9f
Virtual geometry, 32
Virtual material-geometry environment, 208
Virtual models
  actualization process, 48–49
  bidirectional input, 67
  object-oriented approach, 49
Virtual reality (VR) space, usage, 54
Virtual real-ness, 29–30
Visioning, usage, 221
Visualizations, creation, 176

Walk-out basement, permission, 172–173
Watt, James, 82, 224–225

Weight performance, 190–191
Weitzman School of Design
   robot arm, usage, 216f–217f
   Robotic and Additive
     Manufacturing Lab, 160f
West Coast, seismic requirements, 205
WhatsApp, usage, 167
Wholeness, basis, 230
Wire-based foam cutters, usage, 192
Wireframe geometry, naming conventions (application), 44
Witt, Andrew, 216, 218
Wittkower, Ruldolf, 171
Wood-framed/light-gauge modular systems, differences, 181
Workflow, 115, 148
   design workflows, 171
   economic rationality, 101–102
   gaming, 102–114
   issue, 180
   modular materials, relationship, 179–181
   point-line-surface workflow, introduction, 126

Zaha Hadid Architects (ZHA), 230
   aesthetics, 102
   archive/exhibition team, responsibility, 97
   by-project basis, 101
   circular factory role, 98
   clusters, recomposition, 110–111
   configuration/assembly processes, 97–100
   cross-team pollination, 96–97
   customization, usage, 100
   design research, 96–97
   developer space, 102
   housing typologies, issues, 111
   individual user input, visualization, 108
   Instagram account, usage, 102
   integration platform model, usage, 99
   metaverse, workflow (gaming), 102–114
   micro-factory, 106, 108
   mid-block sites, 98–99
   platform design model, 99
   prefabrication components, creation, 97–98
   projects, pipeline, 99
   re-risking architects, impact, 114–116
   scalability, meaning, 113
   solutions, user activation, 105
   supplier scope, 114–116
   sustainability approach, 104
   three-dimensional configurator environment navigation, 109–110
   typology, 99, 113f
   user-based geometries, 112–113
   user-input geometry, usage, 111f
   user interaction, 98–99
   viewing rights, allowance, 98–99
   workflow, economic rationality, 101–102
   work/innovation, 96
   ZHACode, usage, 109
Zaha Hadid Architects (ZHA) configurator
   backend, existence, 113
   deployment, 99
   development, 98
   logic, 101–102
   sessions, running, 108–109
   transparency, impact, 115
Zero-earth strategy, 138–139
Zoom, usage, 167

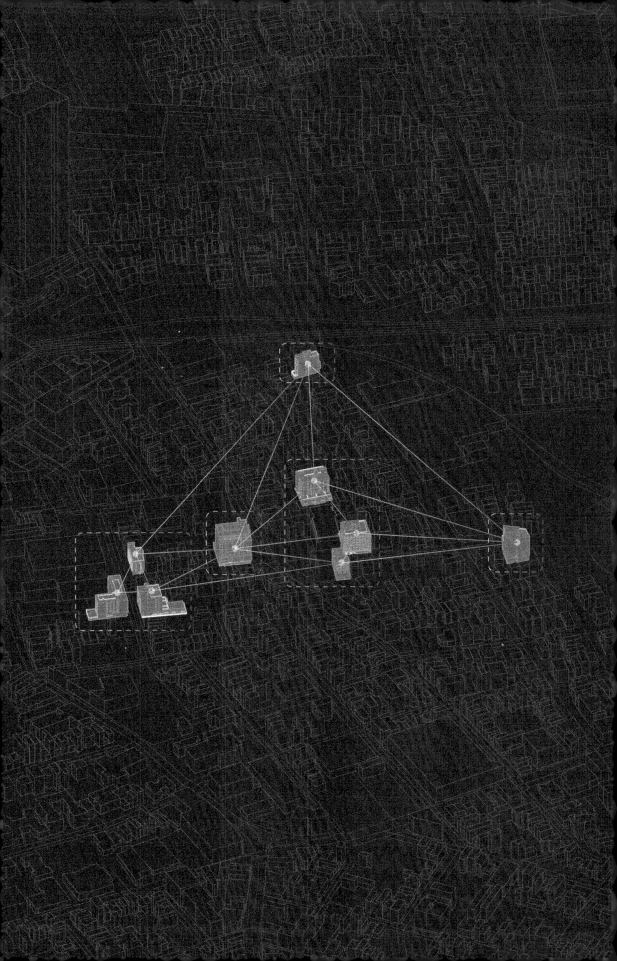